WA 1343

D0247511

STANDARD LOAN
UNIVERSITY OF GLAMORG
TREFOREST LEA

The Agrarian Revolution

John Addy

LONGMAN

LONGMAN GROUP LIMITED
London

*Associated companies, branches and
representatives throughout the world*

© Longman Group Ltd 1972

All rights reserved. No part of this
publication may be reproduced, stored in
a retrieval system, or transmitted in any form
or by any means, electronic, mechanical,
photocopying, recording, or otherwise,
without the prior permission of the
Copyright owner.

First published 1972

ISBN 0 582 31430 5

Printed in Great Britain by
Compton Printing Ltd, Aylesbury

To the memory of
W. E. TATE

Learning Resources
Centre

Contents

Introduction to the Series

The seminar method of teaching is being used increasingly. It is a way of learning in smaller groups through discussion, designed both to get away from and to supplement the basic lecture techniques. To be successful, the members of a seminar must be informed — or else, in the unkind phrase of a cynic — it can be a 'pooling of ignorance'. The chapter in the textbook of English or European history by its nature cannot provide material in this depth, but at the same time the full academic work may be too long and perhaps too advanced.

For this reason we have invited practising teachers to contribute short studies on specialised aspects of British and European history with these special needs in mind. For this series the authors have been asked to provide, in addition to their basic analysis, a full selection of documentary material of all kinds and an up-to-date and comprehensive bibliography. Both these sections are referred to in the text, but it is hoped that they will prove to be valuable teaching and learning aids in themselves.

Note on the System of References:

A bold number in round brackets **(5)** in the text refers the reader to the corresponding entry in the Bibliography section at the end of the book.

A bold number in square brackets, preceded by 'doc' **[docs 6, 8]** refers the reader to the corresponding items in the section of Documents, which follows the main text.

PATRICK RICHARDSON
General Editor

Acknowledgements

The author and publisher are grateful to Leeds City Archives for permission to redraw the section of a plan of Healaugh Manor on page 28 and to the Museum of English Rural Life, University of Reading for the illustration on the cover.

We are grateful to the following for permission to reproduce copyright material:

The Librarian of Barnsley and District Holgate Grammar School for extracts from Enclosure Act for Cumberworth 1799 and Shelley Enclosure Award 1799; Cambridge University Press for an extract from *A Discourse of the Common Weal of this Realm of England 1581* edited by E. Lamond; J. D. Chambers and G. E. Mingay for extracts from *The Agrarian Revolution 1750-1850*; Lord Harewood for an extract from the Steward's letter book to the Earl of Harewood, 14 May 1765 to July 1765, © Lord Harewood; Mr David Hey for an extract from *The 1801 crop returns for South Yorkshire* from Yorkshire Archaeological Society Journal, vol. 42, 1971; The Carlisle Diocesan Registrar for a letter about enclosures, dated 27 September 1821, from Arlecdon parish papers from the Carlisle Diocesan Records, The Record Office, The Castle, Carlisle; The Earl of Mexborough for extracts from the Earl of Mexborough Muniments and the Mexborough Enclosure Dispute 1734; The Controller of Her Majesty's Stationery Office for material from the Public Record Office; Lancashire Record Office for extracts from Scarisbrick Hall Papers and Archdeaconry of Richmond Archives; The Leeds City Librarian and the Leeds City Archivist for extracts from the Vavasour Muniments, the Archdeaconry of Richmond Archives (glebe terriers) and Healaugh manorial maps; Earl Fitzwilliam and his Trustees and the City Librarian of Sheffield City Libraries for material from the Wentworth Woodhouse Muniments; The Earl of Wharncliffe and the City Librarian of Sheffield City Libraries for material from the Wharncliffe Muniments; S. W. Fraser, Esq. and the City Librarian of Sheffield City Libraries for material from the Spencer Stanhope Collection; City Librarian of Sheffield City Libraries for extracts from Rockingham correspondence and Rockingham Papers, Barnsley Enclosure Award and Eccleshall Award; Mrs

Margery Tate for extracts from W. E. Tate's private collection of papers; York Diocesan Registrar and the Archivist for extracts from Tithe Cause Paper G, 1689 and the Archdeacon of York's Court Book YV/CB 1723 from the York Diocesan Records and the Borthwick Institute of Historical Research; Yorkshire Archaeological Society for extracts from Bretton Hall Archives DD 70/60, and Yorkshire Deeds, vol. X, Winterhill Farm Accounts MS 875.

Introduction

In order to study the Agrarian Revolution in its setting of the eighteenth century, it is essential that English farming be seen as a living industry, changing as geography and population made demands on it. Only too often it is assumed that English farming was a static system, of open fields, commons and wastes, until the middle of the eighteenth century when suddenly it was swept away overnight.

This study is an attempt to correct that image by beginning with the early attempts at enclosure and improvement in the thirteenth century. Inflation plays an important part in the changes of the sixteenth and seventeenth centuries while the result of experiment, and demands for more and yet more food lead to the rapid changes of the Agrarian Revolution. Alongside these changes must be studied the effects of prices and profits, the enterprising landlords who invested surplus profits in industrial development, and also that of transport.

The conclusion cannot be left in an unfinished state by assuming that all was over by 1830, the whole perspective is summed up in what is usually referred to as the period of 'High Farming' or the 'Golden Age of the English Farmer'. It is only with the coming of intense foreign competition that farming once again had to change, a study which is beyond the scope of this book.

In selecting the documents for incorporation in this study, attention has been given to those which are relatively unknown but are of importance in the study of this subject. Another reason behind their selection is to open up to the student the large masses of valuable documentary material lying in our county record offices.

PART ONE

The Background

1 Agrarian Development to 1700

Whenever the subject of the Agrarian Revolution is raised, it is commonly assumed that the speaker is referring to the large-scale changes that took place in English farming during the second half of the eighteenth century and the first quarter of the nineteenth. These changes were part of a very long process which stretches back over many centuries and it would be wrong to base this study on the belief that at some point in the eighteenth century the old system of farming gave way to a new one, almost overnight as it were. The truth is that agricultural practice has never been static at any time; change has always been present in the history of farming. Changes in the practice of agriculture have always been accompanied by changes in the arrangement of the land (1). No study of farming can be undertaken without taking into account the part played by geographical factors on the type of agriculture that is practised in any part of the country. In the lowlands of the south and east a great deal of corn was grown but in the hilly districts of the north and Wales, where much of the soil is unsuitable for arable farming, sheep walks and extensive hill grazing existed from early times.

MEDIEVAL FARMING

The traditional accounts of the origin of medieval agriculture — for it is at this point any study of English farming ought to begin — which were accepted by Seebohm, Vinogradoff and Maitland are now discredited. The Anglo-Saxon communities which settled in this island, were practical farmers whose lives were occupied by the grim business of keeping alive, hence their farming practice and land arrangements were the result of commonsense and had nothing to do with political organisation or social status. It should not be automatically assumed that there existed in England a Celtic field system of small rectangular closes, which the Anglo-Saxons took over and replaced later by the oblong strips of the open field system. As both the Celts and the Saxons used the same kind of plough, their methods were controlled by the physical structure of the ground, since ploughing could not be carried out on the hillsides and in narrow valleys exactly as it was on flat ground.

3

One great improvement came in the eleventh century to meet the demand for more food due to an increase, small to us, in the population. The original plough used in Mediterranean lands was light, with a downward pointing sharp spike, which scratched the light, dry soils in these lands to produce an adequate crop. On the heavy clay lands of northern Europe this type of plough was of no use, so only areas of light soil could be cultivated. The new invention was a heavy plough fitted with a ploughshare and coulter to cut deep into the soil, and having a mouldboard to turn the soil to form a ridge and furrow, so draining and digging the ground at the same time. This heavy plough required a team of oxen to draw it across a field, so the tendency was to plough a long strip rather than a square field. This meant that the farmer could now arrange his day's work in a more convenient form. He could plough two contiguous strips, with all the furrows in one strip running in one direction, and all in the other strip in the opposite direction, so giving rise to the 'mingle mangle' of strips described by Tudor writers on farming (32).

In time the length of the furrow became fixed at 220 yards, which was the distance the plough team could travel before needing a rest. The width of two strips of twenty-two yards was determined by the capacity of the plough teams working a full day, which gives the present acre of 4,840 square yards. All the strips did not work out at an acre in extent because the day's work depended on the condition of the land, whether this was flat or sloping and the soil dry and porous or wet and clayey. So the acre was merely a conventional term for a day's work and not a unit of measurement. The acre plots appear to have been handed out in rotation, which resulted in the patchwork appearance of the land.

Most medieval villages could show their 'open fields' which stood in distinction to the closes and assarts that were to be found in every part of the country. Outside these were the meadow lands and the common waste, which was used as pasture. This open field system was not found in Kent, nor in the east, nor along the west side of Devon to Lancashire. The pastoral village of isolated homesteads, scattered over the country, rather than the nucleated agricultural villages, predominated in Wales and northern England because the land was more suitable for stock raising than for arable cultivation.

It is not the purpose of this study to enlarge on this introduction to medieval farming, but rather to show the development of conditions and the mounting social pressures which culminated in the huge enclosure movement of the eighteenth century. This movement was coupled with developments in scientific farming and called for new techniques. In the thirteenth century, the population began to increase, and as villages were small, it became possible to enclose additional land from the waste, since the waste land of a village was not clearly divided from the rest. This type of enclosure was carried out with the lord's permission and was known as assarting. This was a process by which a villein or his younger sons cleared a portion of the waste, fenced it off from the rest, grew those crops he desired and which the land would produce best, paying to the lord a small rent in return. The surplus produce grown on this land was sold in the open market. These enclosures were known by the various names of assarts, intakes, brecks, or new fields. In due course the leasing of these assarts often accompanied the leasing of other lands, examples of which can be observed at Woolley and Skelmanthorpe, in the West Riding of Yorkshire, where waste land was plentiful (62) [docs 1, 2].

This method of obtaining additional land for the production of arable crops satisfied the land hunger of an increasing population. As this increase created a demand for more goods, so a rise in prices took place, and people whose incomes were affected by these increases in prices had to look around for other ways of augmenting their income. Lords of the manor were affected by the price rise for their incomes were relatively fixed, and they saw that one way out of the difficulty was to enclose the waste and the common land and so incorporate it in their demesnes [doc. 3].

The villagers complained when they saw their pasture land dwindling as the result of enclosure. To satisfy their grievances the Statute of Merton 1235 attempted to place restrictions on enclosure by forbidding the lord to take in more of the common pasture than was necessary for the wellbeing of his tenants. The statute did permit the lord to enclose up to the limit but quite frequently when the lord had enclosed and enlarged his demesne, he began to lease it in small farm units. This method helped to increase his income while his tenants benefited by being able to use the additional land to grow more food. The steady increase in prosperity came to an end in 1316, following a series of wet summers which brought in their train both malnutrition and disease. The outbreak of the Hundred Years War drew away the fittest men to fight in France. Those who remained at home were often undernourished and so could make but little resistance to the Black Death which swept across the country between 1348 and 1349. The effect of this was a serious shortage of labour, shortage of food, a general demand for higher wages, and an increase in prices due to shortages in supplies [docs 4, 5].

One important result of the Black Death was a movement of labour. Villeins now tended to move to other manors where wages were higher than in their own. It also gave an opportunity for villeins to move away from poor land to better, and from hard masters to easier ones. The landowners in parliament attempted to check rising costs by passing the Ordinance of Labourers (1349) and the Statute of Labourers (1351). Under these two Acts labourers were to work for the same rates of pay as in 1346, and to stay in their old jobs until their contracts were completed. No alms were to be given to the able-bodied and villeins were not to leave their old manors if there was work for them. Fines were to be levied for breaches of the statute and these were to be used for the relief of badly hit villages. This statute was not an attack on the peasants for according to medieval economic theory, wages, like everything else, had a just level. For anyone to try to raise that level in times of scarcity was an un-Christian act. The statute did attempt to control prices to help the villeins but failed in the long term.

The returns from the East Riding of Yorkshire show that only in a minority of cases did the Black Death empty the villages, but that other factors were involved. For example, there were lords who would break the statute and offer incoming tenants better terms rather than see their land go out of cultivation. The internal movement from one place to another more attractive could lead to the depopulation of a marginal village just as though the plague had killed everyone. The report on the village of Wyvill emphasises this point: 'The three carucates are worth little for the land is poor and stony and lies uncultivated for want of tenants after the pestilence' (*Chancery Cases* C.135/164/7, P.R.O.).

Another solution to the problem was to convert land from arable to pasture for sheep farming. In the fourteenth century there was a big demand for English wool in Europe, for the quality was the finest available anywhere on the continent at that time. This offered a solution to the labour problem, but in some cases it involved a measure of depopulation. This practice of conversion had been carried out earlier by the Cistercian abbeys in creating their grange system of farming. Existing peasant communities which were in the way of progress were transplanted to other sites. For example Rievaulx Abbey moved the entire village of Old Byland from the riverside to a new site on the hill. Scattered holdings were combined and exchange and purchases made to create a new type of farming unit. Fountains Abbey enclosed the tofts and crofts, which belonged to the former inhabitants of Greenberg, destroyed to form their grange (**19**). There is evidence that depopulation was also carried out for the purpose of creating a park, as an enquiry into the village of East Lilling reveals. In 1377 it was a living community but two centuries later the report gives but slender evidence of its existence **(46)** [**doc. 9**].

The Cistercians began, with their grange system, a new type of 'scientific' agriculture. They built great farms on the hills and on the flood plains of the rivers, many of which have survived to the present day. This system led in the fifteenth century to a period of progressive farming that became a turning point in medieval society. The survival of the grange as a unit depended on the profits that it brought to the monastic community. In the north there was a tendency for other factors to militate against it. The growth of the cloth manufacturing industry in the West Riding and the frequent raids by the Scots made many consider whether corn was the best crop to grow. In 1361 Fountains Abbey found it essential to lease their grange at Kirby Wiske and to break up the granges at Marton because they had been so badly devastated by the Scots that the cost of restoration was beyond their finances. Despite this loss their grange lands were very extensive at the dissolution **(46)**.

By the end of the fourteenth century a rapid decline in the manorial system had taken place. Lay landlords were now imitating the monastic communities and cultivating their land by hired labour, while others bought back land to lease at higher rents. Sometimes land was let to a tenant on a stock and land lease. This factor in conjunction with the extension of sheep farming led to a surplus of labour by 1490 **(40)**.

There is evidence that the land on the Yorkshire wolds was becoming worked out, so the plough was withdrawn, and the land reverted to pasture. Many Yorkshire villages which were destroyed by the Scots in the fourteenth century were never resettled. If the fourteenth and fifteenth centuries had been periods in which corn prices were high and the demand for land great then many villages would never have been abandoned for grazing purposes. The cause seems to lie in political trouble existing alongside an economic depression, of which the latter was the more important. These two factors await their historian for we know but little about them as yet.

TUDOR ENCLOSURES

The final phase in agricultural change, before the major developments of the late seventeenth and early eighteenth centuries, occurred in the sixteenth century. The root cause was inflation on a large scale resulting from a combination of several factors. One was the increase in the population which led to an increasing demand for food and goods.

Another was the enormous amount of land speculation which followed the suppression of the monasteries. At the same time there was an increase in the amount of currency in circulation and this expanded at a far greater rate than did the production of food and goods. Thus, with too much money chasing too few goods, there was a fall in the value of European currencies and a corresponding rise in prices. In England the crown did not help matters, for in order to make a profit for a hard pressed treasury it began to carry out a series of devaluations between 1526 and 1551 which considerably reduced the value of English coinage. The effect of this devaluation on landlords was to make them realise that, if they were to maintain their standards of living they would have to find new and more profitable ways of farming their lands. Socially conscious Tudor writers such as More, Latimer, Wilson and Hales criticised the debasement of the currency and those farmers who found a new way of making profit by enclosing for sheep farming (53) [doc. 7]. Land itself was not in short supply, but improved or reclaimed land under cultivation was and therefore rose in price, as did foodstuffs and other materials which the land produced.

Rising prices were not the main factor in agricultural change, for there was the pressure of a rising population. Contemporary writers saw the increase in food prices as the incentive for enclosure, but many had their own private reasons for such action. Sir Charles Hussey, Lord of Honnington, stated that he enclosed his land in order to raise rents and increase his income to pay for expensive lawsuits [docs 8, 9]. Others, such as the Earl of Strafford, realised that a farm under the plough needed almost constant supervision of the labourers if it was to be profitable. On the other hand the inhabitants of Bradford in Yorkshire enclosed Bradford Moor because the population of the town had grown and the ancient cultivated land was no longer sufficient to support them. The Lancashire townships which once formed part of the demesne lands of Sawley Abbey complained when they saw the number of new tenements built on the land since the dissolution and fell out with each other over the division of the commons.

In the sheep-rearing hill country, where sheep walks were widespread and the population thin, rising prices seem to have been a factor of some influence in extending the size of the flocks. On the uplands the large farmer found it easy to enclose large areas of common or to force enclosure of the open fields on the lower slopes, because his tenants were few and the opposition weak. This is the reason why so many small settlements were destroyed, whereas in a large village the opposition would have been too strong. So depopulation occurred on the limestone hills, the wolds and the chalk downs.

The lowland farmer received his incentive to enclose from the high price offered for wool and mutton. For the first thirty to forty years of the sixteenth century the devaluation of the currency abroad gave an impetus to the cloth trade so that everyone wanted to share in it, whether they had experience or not. Although the market for cloth slackened after 1551 it did not make much difference to the upland farmer's income. He had to meet a growing demand for mutton for the townsman's table. In addition he produced from his sheep the fine short wool which was scarce and much in demand, whereas the sheep flocks in the valleys, the fens and marshes produced a coarse wool that was not in great demand. The sheep flock of the Tudor farmer did more than fertilize his fields; it paid his rent, furnished his house and fed and clothed his family.

The concern of the crown in the change of land use was that it led, in this case, to unemployment and depopulation, the latter creating a problem of military security. Its concern in the increase in the number of sheep, with a corresponding reduction in the supply of corn, is reflected in a Bill put through parliament to check the wholesale increase in the number of sheep kept by any one person [**doc. 10**]. Engrossing, or joining of farms together into one, was equally important, for this meant the farmhouses were reduced to cottages or left to decay, resulting in social tension. Much of the Tudor legislation was selective in its application and as the demand for wool declined after 1551 so it once again became profitable to grow grain, and by 1593 there was a surplus.

The clergy suffered from the conversion of arable to pasture, for the tithes on wool were of far less value than those on grain crops. When grain prices soared the tithe owners were less anxious to commute their tithes for a *modus decimandi*, whereas the tithe payers saw more profit for themselves if this were possible. Hence various excuses were brought forward to justify the practice and the number of disputes before the ecclesiastical courts and the Exchequer increased [**doc. 11**]. The outbursts of Sir Thomas More and Bishop Latimer against enclosures must be taken for what they are — exaggerated accounts of a process that was by no means so universal or so wholesale in its destructive effects (**49, 51**).

Christopher Hill has estimated that the cost of living rose by 650 per cent between 1500 and 1640, though the increase was by no means even in its rate of growth. The increase between 1593 and 1599 shows that these years were fairly stable so far as prices were concerned. Nevertheless taking all factors into consideration the total increase in prices between 1500 and 1640 was extremely great (**33**). It was the combination of all the factors under review that led to the beginning of the Agrarian Revolution after 1660.

The old system could not meet the demand for more food for a growing population and so in the sixteenth century changes in the working of open field farming began to be made. In 1577 the Cudworths of Thurgoland, near Sheffield, were encroaching on the common waste and enclosing ground into their farmland (39) [doc. 12]. The inhabitants of Cawthorne had ended the system of open field farming in their village by 1614 and were encroaching on the wastes of the next village of Hoyland Swaine. Sixty years later the village of Skelmanthorpe, near Huddersfield, then growing into a manufacturing district, ordered the tenants in the four common fields and the occupiers of strips in the three open fields to divide them up and enclose them by Lady Day 1640 [doc. 13].

The ideas of the earlier writers on agriculture were later put into practice by eighteenth century improvers. The individual farmers and partners in village farms had been unable to experiment because scientific ideas were often mixed with quackery. Mistakes in farming, as the result of scientific experiment, were too costly, especially when the survival of the community was at stake. The causes of the upheaval in sixteenth century agriculture were the price rise, accompanied by a large demand for English wool, and the land speculation which came about as the result of the dissolution of the monasteries. The trade in land was accompanied by enclosure for both arable and pasture. Rents were raised and so were fines on entry to land, both associated with eviction and the introduction of leasehold lands. By the turn of the century there had arisen a considerable class of landless men who earned a living by working as labourers for wages on a day to day basis, or as some did, went into the towns, or, quite commonly, turned to vagabondage and organised crime. A new attitude arose towards land, which was looked on as a source of wealth and not as a basis of political power or social standing in county society.

Tudor agricultural changes initiated the move towards commercial farming and the slow destruction of the English peasant in the south, if not in the north. By 1600 the foundation of modern agricultural society had been laid — that of landlord, leasehold farmer and landless labourer, a relationship which marks the rural scene from that day to this (3). The large redistribution of property also marked the rise of a middle class which came to stand between the crown, the great nobles and the small farmers. What was really new in the social scene by 1600 was the increasing number of gentlemen and their power in relation to other classes in society.

The first agricultural writer of note was Anthony Fitzherbert, whose *Boke of Husbondrye* was printed in 1523. After an experience of forty years as a farmer, Fitzherbert could insist on farms being in individual ownership with the fields divided by hedges and ditches into separate enclosures, because this method gave a longer season of grass, which benefited both animals and corn. He advocated mixed farming, and wrote that, 'a full bullock yard and a full fold make a full stack yard'. Had his advice been adopted between 1480 and 1640 much of the misery caused by converting arable farms to sheep walks would have been mitigated (3) [doc. 14].

Fifty years after Fitzherbert came Thomas Tusser whose book, *Hundreth Good Pointes of Husbandrie* (1557), republished as *Five Hundreth Pointes of Good Husbandrie* (1573) was written in doggerel verse. Tusser, a recorder rather than an improver, advocated the superiority of enclosed land over communally occupied land. He was the enemy of pigeons, rooks and crows, because they ate grain rather than grubs. He encouraged the growing of 'buck-wheat' as green manure, to be sown in May and ploughed in, in July (11).

These two writers were Englishmen without any knowledge of continental practice. Their contemporary, Barnaby Googe, had experience of farming in Holland, which at that time was the most advanced agricultural country in Europe. Googe recommended the use of rapeseed and also of trefoil as the best food for livestock, to which he added turnips, then used as a cattle food in Holland. In England, turnips were regarded as a garden vegetable suitable to be 'boyled and eaten with flesshe'. Clover and turnips were to be new sources of wealth to individual farmers and the pivot of mixed farming in the future. It was impossible to grow turnips on village farms until all the partners could agree to change their system of crop rotation, but many independent farmers were equally backward in this respect.

Sir Richard Weston in 1645 described the cultivation of turnips and artificial grasses in Brabant and Flanders, but in England the times were unfavourable to progress for the country was troubled by the Civil War and political strife. His ideas were not helped by foolish writers such as Adolphus Speed (1659), who recommended turnips to farmers as food for hounds, an ingredient in bread, and producing 'exceeding good Oyl, excellent Syder and yielding two very good crops each year'. Arthur Yarranton succeeded in growing grasses and turnips in Worcestershire between 1653 and 1677, but it was long before they came into common use. Poor communications and disputes with the clergy and lay rectors over the possibilities of claiming crops of turnips as titheable delayed the expansion of this crop, but they seem to have been grown as a field crop at Coverham in north Yorkshire as early as 1695.

In 1600 Sir Hugh Plat stumbled by accident on the advantage of drilling corn. 'A silly wench' dropped wheat seeds into holes made for carrots, and he discovered the yield per acre rose from four to fifteen quarters. As a result, he and his successors dibbled beans as their models, and in 1601 Francis Maxey actually produced a machine to punch holes in the ground.

Fertilisers also attracted attention. Gabriel Potts advocated using 'coloured' water from flooded land, the soil of streams, malt dust, the entrails of animals and blood offal. One contemporary, Walter Blith, included chalk, lime, farmyard manure, ashes, soot, coarse wool clippings, horn, seaweed, bones, fish and urine as suitable fertilisers.

All these ideas could bear little fruit outside the immediate locality in which they were practised until communications improved. The Act of 1555, which laid the burden for the maintenance of roads on the parish, by voluntary labour, resulted in very inadequate repairs undertaken to already very poor roads. The increase in heavy transport in the eighteenth century made these bad parish roads even worse, and for all practical purposes, impassable for carts and waggons, especially in winter. Pack horse trains were expensive but were often the only means of transporting goods. Daniel Defoe found travelling difficult in 1724 (2) [doc. 15]. Arthur Young when travelling from Preston to Wigan in 1778 described in vivid language the terrible state of the main roads in Lancashire (13) [doc. 16].

Until the standard of transport improved any further advance in farming was of little use, for no farmer could produce perishable goods for distant markets. River transport had been advocated by John Taylor in 1634, and by 1700 a number of rivers had been deepened to take barges (60). The steps taken to make rivers more navigable, so characteristic of the early eighteenth century, had been clearly taken with markets for agricultural produce in mind, as the success of the River Wey Navigation to Guildford in opening up the corn-growing country round Farnham had shown.

The importance of canals in promoting agricultural change is perhaps underestimated. In the south and south-west canals were built with agriculture intended as the chief beneficiary. Finance was provided for these canals largely by landowners, and much of the canal 'boom' in the south took place in response to the buoyant food prices during the French Revolutionary and Napoleonic wars. The Kennet and Avon Canal, built by Rennie for £950,000 between 1794 and 1810, was such a project. In 1795 the neighbouring Wilts and Berks Canal was built to carry coal to the Vale of the White Horse and to bring back agricultural produce. In 1838 this canal was carrying 62,000 tons of goods, and was in use until 1914. Almost exclusively agricultural was the Basingstoke Canal, the stated aims of which were to introduce to the countryside the newer techniques of agriculture. Manure, that essence of farming, was to be carried from the chalklands at one penny per ton, compared with one shilling per wagonload. The barges were to return with peat and 'peat ash', lime and sea sand to make areas such as Bagshot Heath cultivable. Similarly the Bude Canal in Cornwall incorporated elaborate engineering to carry loads of sea sand to the hill farms. The benefits of this sort of carriage can be appreciated when it is realised that it took four packhorses working four days each week to carry as much sand as one small barge.

It is less easy to isolate the advantages to agriculture of the more industrial canals, but it is certain they were used this way. The construction of the Aire and Calder Navigation between 1699 and 1703 to connect Leeds and Wakefield with Goole and Hull was for the dual purpose of serving industry and agriculture. The Marquis of Rockingham's development of the Ouse and Derwent Navigation was to open the East Riding to agricultural expansion. Telford, in outlining the advantages of the proposed Union Canal in 1804, included the following items: 'for . . . conveying Manure for the purpose of agriculture; transporting the produce of the Districts, through which the Canal passes, to the different markets; and promoting Agricultural Purposes in general'. The Union Canal was primarily industrial, so that it can be safely assumed that all canals produced these side benefits for farming (**23, 48**).

PART TWO

Analysis

2 Improvements in Scientific Farming Practice

All the problems which appeared in earlier centuries coalesced in the eighteenth. Inflation, the effects of the French wars, increasing population, the growth of industry, the demand for more food, the necessity to take in all available waste land for arable purposes, compelled farmers and landowners to find improved ways of producing crops and animals. With these in mind we pass on to study the changes commonly known as the Agrarian Revolution.

The prelude to the wholesale enclosure of the remaining open fields and commons was the improvement in the rotation of crops, including roots, legumes and grasses, which enabled the land to carry more stock, thus in turn providing more manure to enrich the soil. The real breakthrough in scientific farming was the adoption of two alternative systems of agriculture – the 'Norfolk system' used on the lighter soils, and 'ley farming' or convertible husbandry on heavier soils. Both systems implied a well drained soil being available. At the same time other changes were taking place, such as the improvement of stock by selective breeding, experiments in drainage, the treatment of soils with marl, chalk, bones and other materials, the gradual introduction of new machines and the appearance of better designed farm implements.

The cultivation of turnips, swedes and mangolds, together with rich legumes and grasses, including clover, sainfoin, and rye grass, raised the level of animal nutrition. Improved output and better quality of meat and dairy produce followed. The traditional autumn slaughter of cattle, so often presented as the standard practice, never appears to have been as heavy as has been believed, nevertheless, milk yields were low, so that fattening animals for market was a costly proposition and their potential fertility was never achieved. The practice of mixed grazing of flocks and herds on ill-drained commons, hill pastures and open fields enabled cattle disease, foot rot, and other equally destructive infections to spread easily. The study of tithe cause papers reveals the extent of disease reflected in the low yield of milk from cows.

The turnip was well known as a field crop in Norfolk, Suffolk, the south-east and parts of southern England by the eighteenth century. Reference has already been made to the growing of turnips in north

17

Yorkshire early in the century. Clover, sainfoin, trefoil and lucerne together with the cultivated grasses, such as rye grass, cocksfoot and meadow fescue, were used for grazing stock and as winter hay fodder. The farmer was provided with cash crops of barley, for the brewing of ale and wheat, while grasses, which were turned into hay and turnips bridged the 'hungry gap' of March and April. This was important when a cold spring delayed the growth of pasture grass. No eighteenth century farmer understood the power of clover to fix atmospheric nitrogen, but the fact was grasped that in some way it restored soil fertility after exhaustion by crops of wheat and barley. It was also found that land benefited from ploughing in of leys that were no longer productive. Those who knew about the work of Jethro Tull realised that the systematic hoeing of roots sown in drills provided a fine tilth which helped young plants to grow and kept weeds down. Under the new system land was always growing something of value, either corn for cash or fodder for fattening animals.

The pioneers Tull, Townshend and Coke in the light of recent research (21) appear to have been popularisers of the new methods of farming rather than introducers. They drew on the earlier practices of the Dutch and the ideas of writers such as Richard Weston. Certainly all three seem to have developed existing practices rather than to have transformed their estates. By irrigation, under-drainage, new manures, and with the holding of private agricultural shows, their main contribution was to spread the understanding of crop rotation (8, 66). Between 1778 and 1821 visitors came from all parts of the country to see Coke's farming methods and over 7,000 visitors are said to have attended his 'Holkham Sheep Shearings'; the pioneers of the agricultural show of the present day. The Duke of Bedford and Lord Leconfield both copied Coke's sheep shearings and attracted visitors from a wide area. King George III had a model farm at Windsor and copied the methods advocated by the developers.

The improved cropping system was not adopted everywhere because of the conservative attitudes of farmers and the resistance to new ideas which was often the result of traditional restrictions on experiments and the unsuitable small holdings of open field farming for such development. Eighteenth century writers attacked the small farmers as being responsible for the slow growth of improved farming. Edward Laurence, John Arbuthnot, Arthur Young and the Board of Agriculture's county reporters were among those who believed that the small farmer with his ignorance, traditional outlook and lack of capital was always a bad farmer. William Marshall thought that improvement was less evident in the west and north than in East Anglia and the Midlands (74).

All these reporters forgot the existence of geographical factors which influence farming. There is a difference between the free draining, light sandy soils of Norfolk, the chalk and limestone of the Downs and Cotswolds, and the heavy wet soils of the Midland plain and the clay vales (70). E. L. Jones points out that in Hampshire the growing of new crops enabled stocks to be reduced and arable farming extended (38). W. G. Hoskins notes that in the Vale of Belvoir the rich open field arable was put down to grass and cultivation extended to the poorer land on the hill slopes (36). The whole area of crop production expanded in the traditional granary of the clay vales. Good crops could be grown in a wet season on light soils. The wet, cold clay lands presented a problem when attempting to grow legumes and grasses with current methods of drainage which were ineffective in improving soil adaptability. It would appear that two crops and a fallow were the basic rotation even after enclosure until the introduction of cheap underdrainage in 1845. R. H. Slicker von Bath has described how the idea of intensive farming spread from Flanders to Britain where British farmers found themselves faced with similar problems and wished to farm for an expanding market (57). The new systems could only be successfully exploited on enclosed and fairly large farms.

The development of progressive farming in an expanding industrial region such as South Yorkshire can be observed by a study of the crop returns from 53 parishes, totalling 89,000 acres, in the rural deanery of Doncaster for 1795. These show the chief crop was wheat, followed by oats and barley and generally turnips and potatoes. Roughly one third of the land was arable with a great deal of grassland [doc. 17]. Arthur Young's belief that small farms were of little use is not borne out by a study of the land utilisation (68). A considerable part of the region was farmed by small proprietors who were known to be numerous and useful.

The Norfolk system relied on an alternative use of fodder and corn crops in order to abolish fallow land, to keep the soil in condition and to provide food for animals. In ley farming there were no regular breaks for root crops. Clover, grass and sanfoin were kept down for a number of years until the land was ready to bear three or four consecutive corn crops. Even under the Norfolk system, clover was often left down for two or more years as a ley crop. Both systems were variations of a crop rotation adapted to particular soils and climatic conditions. William Marshall explained this convertible system of husbandry in his report on farming in the Midland counties. (6) [doc. 19].

The great Norfolk farmers were not the only ones to experiment with new methods. Arthur Young, during his great Northern Tour, called at Wentworth Woodhouse in south Yorkshire to see the Marquis of Rockingham and commented favourably on what he saw (14) [doc. 20]. Fortunately for us, the Marquis of Rockingham recorded the details of his experiments in a memorandum book, with comments on their success or failure. He brought to his south Yorkshire estates a Kentish farmer and a farmer from Hertford, with the idea of comparing the Kentish and Hertford systems alongside the local one. The local cycle was of three years, involving a fallow field for two years in three. In the first year turnips, barley, clover and wheat were sown. The second year was fallow, wheat, beans and barley while the third year had a course of fallow, wheat, beans and barley. By contrast, the Hertford system had three courses, which were all alternatives: turnips, barley, clover, wheat, oats, with the second as fallow, wheat and oats, while the third was fallow, wheat, beans and wheat. Rotation on the Kentish system seems to have been on an eight-year cycle: beans, wheat, pease, wheat, turnips or cabbage, barley, clover, wheat. In the long run the Marquis adopted the system which was best suited to the land under the plough in a specific part of his estates and we find in his memorandum book of 1753 the details of the crops grown in that year [doc. 21].

Closely associated with both forms of farming was the improvement of the soil by marling, manuring and draining (72). Marl was a mixture of either clay and lime or chalk and lime added to the soil for a specific purpose. Hence when clay marl was added to a sandy soil it enabled the soil to retain water, so retaining the manure, which was spread on afterwards close to the surface. The addition of chalk marl to a heavy clay soil made it easier to work and assisted drainage. These marls had been known to earlier generations, but for some reason they appear to have been discontinued except in Cheshire, Lancashire and Staffordshire. In the eighteenth century the sandy soils of Norfolk and Suffolk were marled with chalk brought from the Kent quarries. Daniel Defoe noticed the importance of this when he described the transport of chalk from Kent to East Anglia (2). A study of eighteenth century leases reveals that landowners laid down detailed instructions as to the number of acres that had to be marled each year during the whole term of the lease. Sir Thomas Blackett of Bretton Hall, Wakefield, had such conditions written into the leases of his farms in High Hoyland. Many varieties of other types of fertilisers, such as pigeons' dung, tan bark (from tanpits), saltpetre, oxblood, soot and bones were used. All these appear to have been tried out by the Marquis of Rockingham on his estates [doc. 22].

The most precious material was farmyard manure, and it was considered a crime to sell this or to put it on the land in a lump. Inventories attached to the wills of farmers in the eighteenth century and earlier all contain a valuation of the manure in the farmyard. The burning of lime was widely practised, and the Devonshire custom of stripping turf, burning it and spreading the ashes on the land to clean the soil, as well as enrich it, was well known. Waste materials, from local industrial processes could be also used as fertilisers. Such items were the parings of horn from shaping knife handles in Sheffield and the shoddy waste from the woollen mills of Dewsbury and Batley. All these methods can be found in use on the Rockingham estates [**doc. 22**].

Drainage represents a similar problem. The ridge and furrow method on heavy soil was the commonest; in wet districts of the country the depth from the top of the ridge to the bottom of the furrow could be as much as three feet. This system was wasteful, for soil was washed away from the ridge and water stood in the furrow making it a ditch. Some experiments with under-drainage were made, the most popular being trenches filled with bracken or furze and the soil being replaced above it. In East Anglia wedge-shaped drains were constructed filled with twisted straw, or sometimes soak pits filled with boughs of ash or alder were made at intervals. The Wentworth estates appear to have used stone drains and some of these original ones have been uncovered during recent ploughing of old pastures at Spring Wood Farm and Upper Hoyland Farm, near Barnsley. Recent open cast mining activities in the Barnsley area have brought to light a number of stone drains about ten to twelve inches wide and ten inches deep, covered with heavy top flagstones to bear the weight of the soil and prevent damage by the plough. Incidentally these drains were still functioning [**doc. 22**].

It was the early nineteenth century before real progress in this field was made. In 1823 James Smith of Deanstone, Perthshire, cut a series of trenches which he filled with stones to act as a drain. The results of his experiments he published in 1831 in a book, *Remarks on Thorough Drainage and Deep Ploughing*. When John Reade produced a cylindrical clay pipe or land tile it made possible the more efficient draining of clay soils. The invention by Thomas Scraggs in 1845 of a machine to mass-produce these land tiles made drainage cheaper and gave farmers on clay soils a longer season.

Attempts to improve breeds of animals had been tried for some considerable time. By the eighteenth century Welsh and Scottish cattle were useful for their beef and milk. It was discovered that red or brown Midland cattle when crossed with Dutch stock gave a better milk yield.

From Lincoln northwards were found Yorkshire and Durham short-horns. Sheep were of four types — the middlewool and shortwool of south-west England and the Midlands, while in the north were the horned black-faced sheep and the horned Scottish sheep, with the long-wools of Lincoln, Leicester, Kent and the Cotswolds. Eighteenth century improvers were more concerned with the quality of mutton than wool, and, as the feeding of sheep on clover coarsened the fleece, this led to a demand for imported Irish and Spanish merino sheep (45). George III was most interested in this aspect of improving fleeces that he offered Sir James Eccleston of Lancashire one of his Spanish merino rams to improve his flock [doc. 23].

Robert Bakewell began his experiments in 1745, although he was but one out of fifteen or twenty men who were concerned with animal improvement. His fame arose from the fact that he was a specialist pedigree breeder, who obtained publicity through the writings of Arthur Young. Bakewell began with Lincoln sheep crossed with Rye-land to produce the New Leicesters which fattened quickly, so pro-ducing the fat mutton which was popular with the poor. The New Leicesters improved the breed of Wensleydale, Lincoln, Leicester cross-breds and Penistone sheep. Bishop Richard Watson of Llandaff con-sulted Bakewell about his meadows, watered with irrigation channels, on his estate in Westmorland. Through Arthur Young, Bakewell received a great deal of publicity and many visitors. '11 June 1787; Earl of Hopetown and his Agent, Mr Steward breakfasted with me on Monday. 16 June; Marquis of Graham at Dishley from 11 to 5 o'clock'.

It was not only Bakewell who was active in this field, but also Webster of Canley and Foster of Rollright, whose longhorns were better than Bakewell's. In 1780 John Ellman of Glynde, Sussex, improved his shortwool sheep, and, together with Jonas Webb, transformed the Southdown from a light long-legged animal into a solid compact one for mutton and wool. In 1787 the Colling Brothers of Durham, the Booths and Thomas Bates were busy improving shorthorn cattle which replaced Bakewell's longhorns and laid the foundations of pedigree breeding. Matthew and George Culley of Denton, Darlington, introduced Bake-well's Leicestershire breed of sheep into the north, crossing it with the Teeswater ewe to produce the breed later known as the Border Leicester. They also developed the shorthorn breed of cattle, and by selective crossing, improved feeding methods, cultivation of turnips and attention to soil improvement were able to produce a fatter stock in a shorter time than had previously been possible (67). On heavy clay soil, cattle and not horses were used for the plough. The Marquis of

Rockingham found that by buying small Welsh oxen he could use them for two purposes.

> Sir Robert [Walpole] informed me Bullocks were much more profitable for Teams employed at Home than Horses. *N.B.* A Pair of Great Oxen cost £24. Oxen for Teams bought in Wales at or near £8 the Yoak & will be more profitable for me as becoming afterwards good sized Beef for my own Table & being small more properer to be fed on the Land about Wentworth.

So wrote the Marquis in his Memorandum Book.

The size and style of farm implements varied from area to area and a wide variety of designs persisted until the twentieth century. The size of farms, the varied conditions and supply of cheap labour were the factors which influenced the slow rate of change. In 1730 the Rotherham plough came to fame in the north and east. This was a swing plough with a curved mouldboard which formed a better furrow than the straight board. The design of the main frame was small and lighter so requiring fewer animals to pull it across the fields (26). In the 1780s

The Rotherham plough

Robert Ransome, the founder of the firm of Ransome & Rapier of Ipswich, introduced his self-sharpening, hardened cast iron ploughshare. He also developed the cast iron land-roller and harrow which replaced the medieval log or hawthorn tree tied to a frame.

A century after Tull had invented the seed drill, Arthur Young could find only twelve farmers using it in Hertford, although the Marquis of Rockingham was using both the hoe and drill on his land in South Yorkshire. 'They have also tried & find the Drill Husbandry succeeds extremely well for Poor Land & in Beans and Peas'. Mowing and reaping were amongst the tedious tasks of the farm and also two of the hardest. They consumed a large amount of labour so that all the farmer's family and any local craftsmen and industrial workers were called in to help. Colliery leases often had written into them conditions that the miners should cease to work coal and help with the harvest [doc. 24]. The problem of taking advantage of suitable weather in exposed moorland parishes at harvest time was solved in various ways. At Bradfield, for example, the vicar cancelled evensong on one Sunday in September 1723 and took the congregation to help gather in his barley [doc. 25]. Wandering Irish and Lancashire labourers were also employed. Arthur Young described the whole process of harvesting which was slow, tedious and laborious (13). Occasionally one person would organise a harvest team and accept responsibility for its remuneration. The farm accounts of the Elmhirst family at Worsborough, Barnsley, for 1825-26 show how slow this process was and what a long time it took before the new invention of the mechanical threshing machine spread through the country (70) [doc. 26].

The deepening of rivers, the construction of canals, and improved roads made the transport of goods easier and in larger quantities than could be transported by pack-horse or farm cart. Larger carts were constructed to transport bulk loads to the canal or river which made it possible, by the use of barges, to transport perishable goods to a wider market in large quantities. These latter improvements were incentives to additional production and not in themselves the cause of agricultural change.

Another feature of eighteenth century agriculture was the printing of books on farming practice at cheap prices. Books were now read more widely as literacy amongst farmers increased. In 1771 Jethro Tull published his book on *Horse Hoeing Husbandry*, which gave advice to farmers concerning the working of soil round plants to let in air. The most prolific writer was Arthur Young who wrote several books. In 1770 he published *A Course of Experimental Agriculture*, and this was followed by the publication, between 1768 and 1771 of his tours, *A Six Weeks Tour through the Southern Counties of England and Wales, A Six Months Tour through the North of England* and *A Farmer's Tour through the East of England*, all of which described in detail the farming customs which he found. From 1784 onwards he edited a periodical, *The Annals of Agriculture*, which ran to forty-five volumes. William Marshall was one of the reporters to the Board of Agriculture, commenting on what he found as he visited different areas of the country. John Spencer of Cannon Hall, Barnsley, wrote a *Catechism for Farmers*, consisting of question and answer which was used in South Yorkshire. The reports made to the Board of Agriculture are of vital importance to the student of this period in history.

3 Eighteenth Century Enclosure

It is commonly assumed that the main obstacle to improvements in agriculture was the open field system of farming. Arthur Young condemned wholesale everything about the open fields describing the farmers as 'goths and vandals'. Reporting for the Board of Agriculture on Oxfordshire he described the gulf separating the progressive farmer on enclosed lands from the conservative diehard of open field farming.

Certainly there were disadvantages to the open fields, such as dispersed holdings, loss of time in travelling from strip to strip with implements, a waste of land in balks (yet these could be used for pasture), a rigid rotation of two crops and a fallow, and common herding, which encouraged disease and discouraged improved breeding. The major weakness was the annual fallowing of one quarter to a third of the arable land in order to restore its fertility, and successive ploughing of this fallow in order to eliminate weeds.

Open field farming was not entirely static, for many changes took place which showed its vitality. In many villages there was a growth of small enclosures, an increase in the number of fields, a greater freedom in crop rotation and in the cultivation of the fallow field. Professor Tawney's statement that the open fields were 'a perverse miracle of squalid putrefaction' and their farmers 'the slaves of organised torpor' is no longer tenable (81). Open field farming could be made much more adaptable and efficient if the open arable fields were supported by sufficient enclosed pasture for the rearing and fattening of cattle as well as dairy farming. There was a strong tendency for these valuable closes to grow in number due to the low corn prices in the early part of the eighteenth century, when farmers agreed to take land out of the open fields for pasture. This happened on the Duke of Kingston's estates between 1730 and 1740, while at Eakring in Nottinghamshire some thirty-five acres were taken out of the grass field between 1744 and 1746 (49). Hence the open fields tended to shrink in size.

Enclosure was taking place at Mexborough between 1688 and 1736 in this piecemeal fashion, and a study of the map reveals that the old field system could easily be reconstructed [doc. 27]. Often exchanges of land took place to obtain more compact holdings and to allow more advanced cropping (32). Farmers cultivating open fields reduced fallowing by growing sainfoin and later turnips; sainfoin fed larger flocks and herds so that more manure was available. By the extension of closes and open field leys, farmers could overcome the shortage of pasture, which formerly limited the number of animals they could keep, and also maintain the reduced amount of arable land in better shape. W. G. Hoskins found that the 'evidence of a convertible husbandry in the open fields goes back into the sixteenth century and probably earlier than that' (34). Therefore enclosure is not the result of open field farming, neither is it a sudden break with the past and tradition, but the speeding up of a gradual process either by mutual agreement or an Act of Parliament. It brought a large increase in the cultivated area and in the productivity of the soil as a response to prices and market conditions. The best use of the land could only be obtained in enclosed and fairly large farms.

The enclosure movement affected some 6 million acres or 25 per cent of the arable land and was most heavily concentrated in periods when England was occupied with foreign wars. In the two periods 1760 to 1780 and 1792 to 1815 some 900 enclosure Acts in the first period, and 2,000 in the latter were passed by Parliament. In the eighteenth century enclosure was most heavily concentrated within an area of the country bounded by Yorkshire on the north, by a line drawn from the Pennines to Bristol to the west, and on the eastern side by Kent, Surrey, Sussex and Hertfordshire. Within this area lay the remains of the open fields which were enclosed. In Yorkshire, Derbyshire, Lincoln and East Anglia the enclosures were chiefly concerned with commons and waste lands.

A great deal of enclosure took place by private agreement, at least until about 1760, sometimes willingly and occasionally under pressure from powerful landowners. A case brought into Star Chamber in 1614 by Sir Francis Wortley concerning trespass by the inhabitants of Cawthorne into his lordship of Hoyland Swaine with intent to acquire additional pasture by moving the boundary stones, reveals that Cawthorne had enclosed all its available land by that date and was in need of more to sustain its population [doc. 28]. Extensive enclosure was taking place in the area of Tankersley between 1593 and 1641, to which William Earl of Strafford (son of the ill-fated Sir Thomas Wentworth) wrote to his cousin Thomas Wentworth of Bretton Hall, Wakefield, asking for advice [doc. 29]. Evidence exists in quantity of small-scale enclosures from open fields and wastes to provide additional closes, the rents having been adjusted accordingly and conditions attached as at Mexborough in 1736 [docs 30, 31].

Two very extensive areas of common were enclosed by private agreement at Brampton and Greaseborough between 1714 and 1727 [**doc. 32**]. These agreements were enrolled in the Chancery Court to give them greater force in law. The methods by which the open fields diminished over the centuries as the result of private enclosure can be observed from a study of the map of Platts Common prepared by William Fairbank for Earl Fitzwilliam in 1771. From a study of this map it will be observed that the Lowe Closes, Back Green Closes and Chapel Close are all in the old Lowe Fields and are in large strip form bounded by fences running in the same direction as the old narrow strips in the open land adjoining the closes. These three closes were the direct result of consolidating blocks of strips by agreement. When Fairbank drew another plan of the manor of Hoyland in 1775 neither the common nor the open fields at Platts Common remained, for within four years all the land had been enclosed. A useful comparison can be made by studying the map of enclosure at Healaugh Park, where the old strips are in smaller units than at Platts Common and the strips more varied.

Occasionally, when enclosure by private agreement had taken place on one of two adjoining manors and the second manor was enclosed by Act of Parliament, unsuccessful attempts were made to force the private enclosers to erect fences along the boundary between their lands and that of their neighbours. Such an attempt was made at Wortley in 1714 when the lands in Thurgoland were enclosed (22) [**doc. 33**]. The correspondence that was exchanged in cases of this nature is the only surviving evidence in many cases of private enclosure.

What then was the aim behind enclosure, either privately or by Act of Parliament? It will be well to consider this before discussing parliamentary enclosure; (1) to make for more efficient farming by building compact and large farms with a balance of arable and pasture. Although the geographical factors of the soil plus the conservative attitude of the farmers made little difference to production, yet enclosed farms often paid a higher rent. (2) to convert old arable land, commons, rough hill grazing, and wastes to a more profitable use by ploughing, marling, and adopting a suitable crop rotation. (3) to expand the area of land by bringing commons and wastes under cultivation in unproductive and lightly cultivated areas. (4) to improve efficiency by getting rid of tithes, in a parish where piecemal enclosure had resulted in an untidy system of tithes in a maze of small closes.

In general, villages with good or improvable soil, effective communications and a small number of landowners tended to be the first to enclose their lands. Professor E. C. K. Gonner found that the Midland

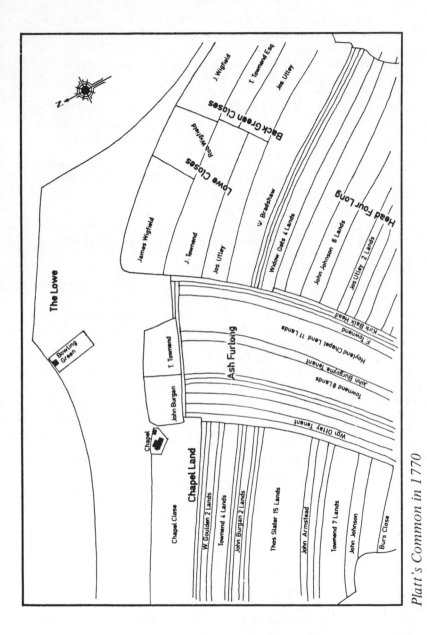

Platt's Common in 1770

Section of Healaugh Manor

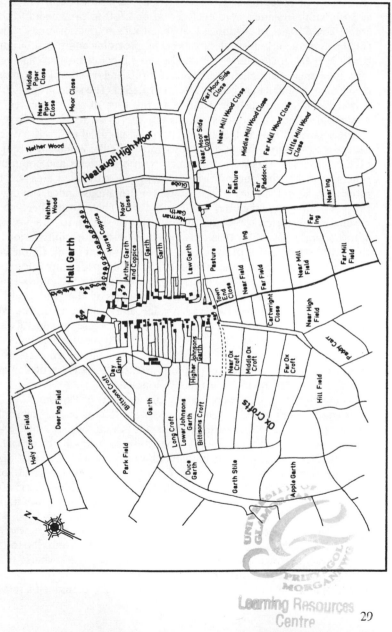

Learning Resources Centre

29

clay lands were enclosed during the early period of enclosure by Act of Parliament in order to convert arable land to pasture (27). The chalk and limestone uplands were enclosed after 1760 to provide additional arable land for arable purposes. Land enclosure in towns after 1770 was for the purpose of industrial expansion by providing additional building land (15). Dr Joan Thirsk found that in Lincolnshire the chalk and limestone areas were enclosed between 1750 and 1780 but the marshland remained unchanged until after 1815 (51). The standards of cultivation varied enormously in relation to the type of soil, the extent of waste and common lands, as well as the size of farms, from village to village. Each village and each enclosure is peculiar to itself, so that generalisation about this is dangerous and difficult. While the majority of the enclosure Acts were concerned with open fields and commons, others were for the purpose of extending the cultivated areas on the moors, while yet others were merely confirmations and a tidying-up process following an earlier enclosure (61). Their study becomes a fascinating historical problem, offering much scope to the younger student working locally. It is only possible to give two illustrations of this within the limits of this book. When the moor at Egton cum Newland in Lancashire was enclosed, the land divided up into closes was to be sold by auction and the money applied to the improvement of the Kent estuary, so that regular flooding of the land could be prevented. At Silkstone, Hoyland Swaine and Cawthorne in the West Riding, three enclosure Acts were required to enclose an area of waste known as Gadding Moor, jointly owned by all three parishes, which is the reason why none of the villages appears on the award map. So in every case there is some item of importance to each particular area which throws light upon its future development.

These enclosure Acts were usually private Acts of Parliament and as such were never printed in the Statutes of the Realm, so the real source of information upon enclosure is in the Commons Journals. Any student concerned with the study of enclosures can easily be misled by being badly informed upon parliamentary procedure. The first stage was the petition to the House of Commons, usually the result of a meeting of interested parties in the local inn. The petition was drafted with the support of the lord of the manor, the incumbent, the impropriator of the tithes, together with the principal proprietors [**doc. 34**]. Under the terms of the General Enclosure Act of 1801 it was not essential to include the name of every landowner, but only the lord of the manor, the owner of the tithes and the large landowners. If the petition was made without the support of the lord of the manor it usually failed. In the Midlands the petition usually contained a request for the commutation of the tithes, as it was at Cumberworth in the West Riding in 1799. After 1774 it became obligatory to give three weeks' notice of enclosure by fixing a notice on the church door for all to see.

The next stage was the ordering and reading of the Bill, and it appears strange to our modern eyes that the members who promoted it were relatives and friends of the petitioners. After the first reading a Committee on the Bill was appointed, which after the second reading set to work. Petitions could be made against the Bill, but, unless the large landowner was concerned, not much notice was taken of them (9). The Bill was then reported with or without amendments for the third reading. The Lords usually passed the Bill as it was or made minor amendments to safeguard the interests of the Church, such as the Easter offering, or correcting a title from Right Honourable to Most Noble, after which it received the royal assent (44, 78).

Commissioners were then appointed to carry out the provisions of the Act. This meant that the commissioners must survey the land before allocating it to the claimants. In practice, surveying of land became a professional occupation for some country gentry, land agents, large farmers, occasionally some of the clergy and, in one case, the fishmonger of Cawood, who acted as enclosure commissioner in York-shire. The skill gained by surveying land for the purpose of enclosure produced in time a group of professional surveyors who were to be valuable when the railway system came to be developed (15). Commissioners had to swear an oath on their appointment that they would act impartially in their task, and a single commissioner could undertake to attend to several enclosures taking place at the same time (76) [doc. 35].

The next stage was to announce in the Press the details of the commissioners' appointment and to notify the village community by fixing a notice to the church door. This notice gave the date and place, usually an inn, of their first meeting, to be followed by the hearing and deciding of claims and counterclaims. Sometimes this took a few months but sometimes, as in the case of Meltham, many years (15) [doc. 36]. During the entire period of enclosure, the commissioners directed the system of agriculture in the open fields, if any were in-volved, while the surveyor drew an exact plan and recorded measure-ments of the lands on which new allotments were to be recorded. Provision had to be made for roads, paths, drainage and a gravel pit or quarry for the purpose of obtaining materials for road building; fences had to be made according to specification. Objections to new roads were frequent, and the new road involved in the Thurgoland enclosure caused much delay [doc. 37].

Occasionally some land was reserved for the poor as cow pastures or as vegetable gardens, or even for the building of a school (10) [doc. 38]. Provision for the loss of common rights and compensation for the tithe owners were made. In some cases, but not in all, tithe in kind was commuted at enclosure on a basis of a corn rent [doc. 39]. Claims had always to be made in writing, and the commissioner had to decide whether or not the evidence was accurate and if the claimant could be given an allotment for his loss of common rights. Frequently, and especially in Yorkshire, as William Marshall pointed out, it was the householders with common rights attached to the tenement rather than the owners of farms who pressed for the enclosure of the common (4) [doc. 40]. Encroachments on the common were to be recognised and compensation awarded if they had existed for twenty-one years or more; those of less standing were to be ignored and the land allotted. The costs of enclosure were by no means light, having to cover the expenses of the commissioners, the cost of the passage of the Bill through Parliament, and finally the cost of fencing the land as specified (78) [doc. 41]. These fencing costs were based on the length of the boundary and not the acreage. A one-acre close has one-tenth of the length of the hundred-acre close. Though the gross cost per acre is one-tenth that of the larger, yet the net cost per acre is ten times greater.

Those who received small plots of land and whose entire holding was tiny found that the expenses involved were too great to make it worthwhile retaining. Usually these people had a second occupation — either weaving, making cutlery, nailmaking, or some other type of domestic industry — so that they sold their land, for there were willing buyers, and entered industry as full-time workers. When all was completed the commissioners, at their final meeting, 'signed, sealed and delivered', the award. The following Sunday at matins in the parish church it was handed to the incumbent to be retained in that church for reference should dispute arise (9).

There is another aspect of enclosure in addition to that of agriculture, and this concerns the enclosure of commons and wastes in the towns. In the case of Barnsley, where the lord of the manor and the impropriator of the tithes were one and the same person — the Duke of Leeds — this was easy to do. In 1777 enclosure of the commons was undertaken and twenty small farms were reduced to four large ones. Pogmoor and the Race Common, on which encroachment had been taking place for some time, were divided up, the reason for this being industrial expansion. A map of 1714 shows that extensive coal mining, both open cast and shallow mine, was taking place on the common. The growth of the linen industry in Barnsley and the expansion of the iron trade made it profitable to enclose open common previously let at 1s. 6d. to 3s. 0d. per acre, to form industrial sites and housing accommodation for workers, which meant the rent value could be increased to 10s. 6d. an acre or more. In Barnsley the enclosure award contains a section dealing exclusively with the preservation of water supplies for the town [doc. 42]. In Sheffield the enclosure of the outlying parts of the town and the town moor followed the rapid encroachment on the common made by the expanding cutlery trade, but in this case there were riots by those who objected to the enclosure of the moor (30, 77). W. G. Hoskins describes vividly the problem faced by Nottingham, which wanted to enclose its common for housing. In this case the 'cowocracy' stopped house building by blocking the enclosure (36).

The Hammonds, in *The Village Labourer* (31), emphasise the opportunities available for the wealthy to obtain their own ends against the wishes of the majority. They argue that the commissioners showed preference in awarding land, ignoring the claims of those who fell foul of the legal niceties involved. The Hammonds quote a number of Acts to show that enclosure was a swindle on a large scale. Professor Gonner, who has examined a far wider range and number of acts, arrives at the conclusion that the commissioners were fair. Some land had more common rights attached to it than other; some land was worth more than other and small farms were rented more highly than large ones. Gonner writes: 'When the gravity and delicacy of the task undertaken by the commissioners is considered the existence of complaint against them is not astonishing. It is a matter for wonder that the complaints are not far louder and more universal' (27). W. E. Tate asserts that the enclosures deliberately aimed at damaging and destroying the small man were most exceptional and not typical of the country in general. He also finds that small proprietors were not opposed to enclosure but in such cases where there was opposition the small farmer did not refuse to sign a petition because he was afraid to oppose the decision of the larger landowners to enclose. According to W. E. Tate, most of the opposition came from large landowners and the clergy who were concerned about their tithes (76) [doc. 43].

However, W. G. Hoskins, working on Leicestershire enclosures, found that at Wigston Magna small farmers had less land after enclosure than before and that engrossing had serious effects on these small farmers (35). In general, parliamentary enclosures seem to have worked fairly as between various classes of proprietors, and in many cases the claims of the squatters and the poor were taken into consideration by the commissioners. It was imperfect justice, but in an age of aristocratic government and exaggerated respect for the rights of property, enclosures were based on custom and equity rather than purely on the letter of the law. Enclosure represented a big step forward towards the recognition of the rights of the small man.

Before making any assessment of the enclosure movement or of the Agrarian Revolution as a whole, a brief reference must be made to such opposition as there was from time to time against the changes. In 1614 there had been a demand for pasture land for animals in Cawthorne where the best land had already been enclosed for arable to feed a growing population. The parishioners of Cawthorne used the Rogation week procession as an excuse to break up the boundary stones and seize for themselves some of the common waste in Hoyland Swaine, the neighbouring village. Sir Francis Wortley, lord of the manor of Hoyland Swaine, had recourse to Star Chamber for restoration of the land (15). Nineteen years later the Hoyland Swaine people themselves turned on their lord, Sir Francis Wortley, for enclosing three acres from their common wastes, which were very extensive, to give a tenant some extra land. They saw in this act the thin end of the wedge and felt that if no protest was made, then all their waste pasture and common would be enclosed and they themselves would suffer. At Mexborough in 1737 the opposition came from two big landowners, William Sylvester and Sir Charles Savile. Sylvester opposed Savile making enclosures in the town fields since these affected his common rights, and when Sylvester made an enclosure then Savile had it destroyed. Sir John Reresby in 1688 had opposed and thrown open the enclosures in these fields, but after his death others then made had been allowed to remain. The real trouble was a shortage of pasture and the commoners intended to take in 'a large parcel of ground belonging to many owners which was designed for an ox pasture because enclosed ground was so scarce for use as pasture' (15). The riots in Sheffield in 1795 were against loss of the town moor and the racecourse. This was the only open land now remaining for the leisure activities of the townsfolk of Sheffield and it is quite clear that the people were determined to frustrate this by force. Commissioners and surveyors were forcibly evicted, windows broken, the jail demolished, and arson practised on selected houses, including Broom Hall, the residence of the vicar. This riot concluded with Sheffield becoming a garrison town for the maintenance of law and order (15).

Lopham in Norfolk was the scene of one of the few instances of violent opposition, where 200 acres of land had been granted to the poor. At the trial four persons were indicted for breaking down fences and levelling the ground, the traditional peasant method of protesting against unjust enclosures. The leader, one Mason, received a sentence four times as heavy as any of the others received. This sentence resembled in severity that passed on the famous character from *Wind in the Willows*, Toad of Toad Hall, for stealing a car **(15)** **[doc. 44]**.

4 Prices and Profits

The Corn Laws always complicated prices for cereals until their repeal in 1846. The purpose of the Corn Laws when they were first introduced was to prevent wide fluctuations in prices by encouraging the export of corn when home prices were low and restricting imports when home prices were high. Amendments to the Corn Laws in 1790 and again in 1815 gave much greater protection to producers, which in turn had disastrous effects on consumers. After all, enclosure was affected by both prices and profits, which in turn were influenced by demand, so it is as well to take these factors into consideration when dealing with the subject of enclosure.

Between 1750 and 1793 prices rose at a more rapid rate than costs, rents, and wages. From 1793 to 1815 taxes increased rapidly to pay for the war and the poor rates rose steeply to subsidise low wages as a result of the Speenhamland system of poor relief. After the end of the war in 1815 prices fell more rapidly than costs so that deflation and bank failures ruined many farmers. By 1840 matters had improved through the reform of the Poor Law in 1834, followed by the Tithe Commutation Act of 1836, coupled with the abandoning of marginal land which had been used for growing crops.

From 1750 to 1790 rents had increased by 40 or 50 per cent with no accompanying large increases in wages or parish rates, so that farmers became more wealthy, often trying to ape the aristocracy (5) [doc. 45]. The outbreak of war in 1793 was accompanied by a swift rise in prices, together with violent changes in farming, with a wild scramble to enclose land while profits were so high. In 1790 wheat cost 48s. to 58s. per quarter. and five years later the price had risen to 90s., but then fell to 58s. a quarter. The autumn of 1800 saw the price of wheat at 113s. 6d. per quarter, and one year later it reached famine price at 119s. 10d. per quarter. The three years from 1810 to 1813 saw an average price for wheat of 112s. a quarter and only for six years between 1793 and 1815 did wheat prices fall to 65s. The rates for barley show a similar fluctuation. Meat prices rose as a result of poor seasons and dear fodder. In 1780 it was possible to buy beef and mutton at around 3s. to 4s. per stone, but by 1802 the price had increased by 70 to 100 per cent. In 1813 meat prices reached their highest level at 8s. 6d. per stone but by 1817 they had fallen to 7s. a stone. These high meat prices played a great part in the Pontefract Election of 1812, when one of the three candidates, Edward Hodgson, bribed the electors with gifts of rabbits, because meat was so expensive at 9d. a pound and meal at 7s. a stone **(15)**.

The weather hit supplies more than reduced imports affected the markets. There were very few good seasons and the poor ones were in long runs. At the same time prices were affected by the financial policy dictated by the Napoleonic Wars which led to inflation and an over-issue of paper notes, rather than by increased freight charges and Napoleon's blockade. The first upswing, resulting from three bad years, was in 1795. Arthur Young, as secretary to the Board of Agriculture, sent out a questionnaire asking for detailed information on the stocks of cereals, a forecast of the crop yield for 1796, rises in wages, the use of substitute foods and prices of fuel, vegetables and other foodstuffs **[doc. 46]**. The year 1795 was also the first year of the operation of the Speenhamland System of poor relief. This system was intended to be a temporary measure for no one saw that these conditions would last for twenty years and that in the end a problem of subsidised labour would arise.

A study of the table of harvest deficiencies between 1794 and 1800 indicates the problems that would arise **(38)** **[doc. 47]**. The effect of these appalling seasons on the prices of wheat, barley and oats in many industrial areas of Yorkshire (Bradford, Sheffield and Leeds) can be observed from the following table of prices of cereals (in quarters).

	Wheat	Barley	Oats
January 1799	49/6	29/4	19/10
June 1800	134/5	69/1	51/1
March 1801	156/2	90/7	47/2

In Yorkshire, the prices of wheat, barley and oats fell considerably in December 1801 as the result of the improved harvest. There were other bad seasons before the end of the war in 1815. Both 1808, 1809, 1811 and 1812 were years with wet summers and in 1812 the harvest again failed. The failure of the harvest in 1799 led to panic conditions and the Duke of Portland sent out a circular to all the bishops in England asking them to obtain information in their dioceses about the crop yield, the market price of foodstuffs, and the use of substitutes **[doc. 48]**. The situation was not helped in any way by the withdrawal of Austria from the war, as the government was compelled to make plans for withdrawal from the coast in case of invasion. In other words, a scorched earth policy had to be planned for all parishes within twelve miles of the sea coast. So details of wagons, horses, mills, ovens and live and dead stock had to be assembled (87). Areas of the country had been threatened by famine in March 1801 when the price of wheat rose to 156s. per quarter. Bread rationing was recommended by Parliament the same year [doc. 49], and in this connection it is interesting to note that the recommended allocation of bread per head was the same as in 1947 when bread was rationed as a result of the war. In some urban areas, barley and potatoes were used to make bread. At Wibsey, near Bradford, many were driven to poverty and riots encouraged by high food prices [doc. 18]. There were riots in Newcastle, Carlisle and other industrial towns when bakers' shops were broken into and the bread seized by rioters (54).

Farmers and landowners, however, enjoyed a long period of prosperity, and land was in great demand. Enclosures were undertaken at a rapid rate, so that the boundaries of cultivation advanced in the moorland areas to a high point which was never attained again. Rents rose as one result of the war but real wages were far behind, so that labourers faced starvation. Goods which cost 5s. in 1790, cost 26s. in 1801 (24). The Speenhamland System did provide a solution − although a defective one − for those areas where wages were inadequate and no alternative means of employment existed.

In the Midlands and the north, expanding industries influenced both the supply of farm labour and the rates of wages, so that there grew up a division between the north and the south. The north tended to become an industrialised area with high wage rates, while the south tended to remain agricultural and an area of low wage rates, at least until after 1850. English farm labour was mobile over short distances only and the working population of the north was recruited from the neighbouring country regions. One effect of high wages in industry was that northern farmers had to compete with industry for labour so wage rates remained at a comparatively high level. On the other hand in the south wage rates were low, especially in rural areas remote from London.

The question of the Corn Laws was irrelevant to arable farms. The acreage had expanded due to war time enclosure and production had been improved by four course rotation coupled with the return of better seasons. After 1819 it was impossible to obtain prices for corn higher than 65s. per quarter and in good years no more than 40s. The average price of wheat between 1820 and 1846 was 57s. 9d. and, although landowners blamed the heavy taxation of the war period and the deflation caused by a return to the gold standard in 1821, they slowly realised that they would have to adjust themselves to realities. Agriculture would not in the future give any yield on capital applied to old cultivated areas or to marginal soils (18). Between 1815 and 1846 the prices of cereals were all above prewar levels and only in four years during this period, did the price fall to the level operating in 1780. The range of prices for farm produce in the Barnsley area may be compared with Cobbett's remarks on low prices (1) [doc. 50].

The wages of agricultural labourers fell fast after 1815 due to demobilisation of the armed forces and widespread unemployment in industry. The general level of weekly wages dropped from between 12 and 15s. to 9 or 10s. in 1822, and rose but little until after the return of better conditions in 1842. Wages were in line with prices but a revival in industry plus a demand for labour in expanding areas brought wages to about 40 per cent above those of 1780. The tax burden was reduced by the abolition of income tax and a change in the malt tax. In the south the poor rates remained severe until the Poor Law Amendment Act of 1834 removed the burden.

PART THREE

Assessment

Learning Resource Centre

5 The Effects of Enclosure and Scientific Farming

In general the Agrarian Revolution affected four classes of society, the landowners, the clergy, the small farmers and the cottagers and squatters. It is the opinion of the Hammonds that enclosure favoured the landowners and the clergy but was fatal to the small farmer, the cottager and the squatter. In order to arrive at a true evaluation it will be as well to study each group in turn. The large estates of the great landowners tended to grow at the expense of the small ones, for there were always those farmers who by inefficiency, extravagance or total inability to attend to affairs had to sell their land to pay their debts. There were also others with wider sources of income, wealth and borrowing powers who were willing to buy. Large owners were always in the market for new property which gave them greater influence locally and enabled them to enjoy political power by exercising controlling influence at elections. Occasionally the purchase was made to extend the boundaries of an estate, to improve the layout of farms, or to facilitate work on the estate (41).

The growth of large estates was in the hands of a class of landlords eager to exploit new sources of revenue and this was a factor in engrossing, enclosure and improved farming of lands. Another result was the increase in tenant farming and the concentration of small tenancies in the hands of the large farmer. There is evidence that the technical improvements were most notable on the large consolidated properties of country gentry and those freeholders who had capital and the interest to carry out experiments. The large tenant farmers could bring enough pressure to bear on landlords to induce them to provide buildings, fences, roads, protection against flooding and some contribution towards the cost of marling the soil.

The large estate was a financial interest, for the enclosed farms yielded a higher rent than the small farms. The profit depended on the ratio between costs and the increase in rent, for rents on the one hand either doubled, trebled or quadrupled, or on the other hand rose only slightly (28) [doc. 51]. The gross return on the investment would be between 15 and 20 per cent, and where large areas of waste had been enclosed the return was considerably higher. This return was far greater than the 5 or 6 per cent which could be obtained from investments in Consols, the Bank of England, or the East India Company [doc. 52]. Agriculture was a better venture for the wealthy than commerce between 1760 and 1815, so there was a keen market for land. Buyers of land were not the big landowners but Indian merchants, country bankers or retired city men who desired the social standing and political influence which land ownership could give them (55). The whole trend was for estates to consist of large farms of 300 acres or more with a nucleus of medium sized ones varying from 100 to 300 acres and this was the typical estate unit over the country. Hence capital was invested in arable land cn a large scale [doc. 53].

The attitude of the clergy towards enclosure changed after 1660. They became anxious to secure for themselves as large a share as possible of the proceeds, for those who advocated enclosure always included the commutation of tithes as one of its advantages. On the whole, but for completely different reasons, the views expressed by William Cobbett on tithe were supported by the clergy: 'As a general Enclosure Bill will pauperise the countryside so a general commutation will lead to the disestablishment of the Church and the abolition of the monarchy' (1). Arthur Young expressed a different view: 'Tithe is the greatest burthen that yet remains on the agriculture of this Kingdom; and if it was universally taken in kind would be sufficient to damp all ideas of improvement' (15).

A detailed examination of the tithe cause papers, either in the diocesan record offices or in the Exchequer Court archives, will reveal how much the farmers objected to giving away one-tenth of any crop increase which their skill and their enterprise made possible. Hence they frequently cheated and annoyed the tithe owner, were he cleric or layman. Tithes were in some cases commuted for cash, but the great tithes of corn, barley and oats were in kind [doc. 54]. Where lay impropriators were also landlords they merely asked the value of the tithe to their rents.

It is difficult to ascertain, until more research has been undertaken in this field, how much clerical opposition to enclosure was due to a disinterested love for the poor or the fear that conversion from arable to pasture might lead to a depreciation in the benefice stipend. Certainly a rector, rather than a vicar, would lose more if his great tithes were reduced or commuted. Catterick's new vicar in 1748, trying to heal the breaches in the parish introduced by his quarrelsome and non-resident predecessor, 'contented himself with a very moderate commutation in lieu of tithes and never . . . cleared more than one hundred and eighty or two hundred pounds per annum by the living; although it has since been raised by some of his successors . . . to upwards of fourteen hundred pounds per annum'. Such increase in the value of many livings was due to imposition of tithe on newly enclosed lands, the revision of an old *modus decimandi* or the ordinary business of enclosure (20). Poor clergy in Yorkshire were assisted by an Enclosure Act of 1713 when the churchwardens of any parish were allowed to enclose up to 60 acres of common and waste and rent it to augment the stipend of the benefice in those cases where it was worth less than £40 a year [doc. 55]. On the whole writers agree that between 1780 and 1830 with a few exceptions the clergy did very well out of enclosures.

The effect on the small farmer, which under this heading includes the owner-occupier, varied considerably. There is a widespread but entirely erroneous belief that before 1750 the great majority farmed for subsistence. It is true that there was a class of smallholders who farmed a few acres to supplement wages and had a very small surplus to exchange for necessities. Those who were farmers proper with 12 or more acres had a surplus, otherwise they could not have purchased stock and seed from the sale of their produce. Further there was a growing non-agricultural population in the towns and a steady expansion of flourishing markets in Barnsley, Leeds, Wakefield and other towns of comparable size. This meant that commercial farming was a reality by 1750.

The Hammonds stated that 'enclosure was fatal to three classes; the small farmer, the cottager and the squatter'. They supposed that small farmers were adversely affected by the costs of enclosure and produced as evidence a few instances where enclosure costs were excessively high, but provided no scrap of evidence to prove that these costs were fatal to the small farmer (31). A closer examination of the evidence shows that it was those with small holdings and not farms, and to whom farming was the second of two occupations, who found the expense was not worth the value of the land. Many such were absentee owners who let out the land to tenants and so were not occupiers. Professor Chambers has pointed out that the expenditure of £50 to £100 on enclosure costs on good land recently enclosed increased the value of the farm by 100 per cent (21). These small owner-occupier farmers had little difficulty in raising a mortgage or selling off half a dozen acres to meet these costs, or in the last resort to sell off all their land at an increased value and become substantial tenant farmers. The estate papers of Bretton Hall or Woolley Hall, Wakefield, are full of such examples, and others of a similar nature can be found throughout the country (42). Some farmers did find themselves unable to redeem their mortgages and sold out.

The Land Tax assessments show that the small owners did not decline on a large scale. They declined in those areas where other factors, such as the heavy burden of the poor rates and the move towards larger farming units, made themselves felt. On the whole the evidence is that there was an increase in the number of smallholders between 1780 and 1815, but the figures are inaccurate in that they give no details of acreage or size of farms. They do show that enclosure was not an important factor in the survival of the small farmer when prices were good, and also that fewest small farmers were found in villages where enclosure took place by agreement (58). The long-term process was for farms to become larger due to specialisation and efficiency, and that large farmers had enough capital to stand the effects of bad seasons and low prices (25). The change was gradual and to speak of the rapid disappearance of the small farmer is incorrect.

In some areas such as Cheshire, Wales and the northern counties, where enclosure was mainly of waste, the numbers of small farmers actually increased. The demand for food and the attraction of good profits played a part in stimulating farmers to spend money and energy on their farms. What really accounted for the survival of the small farmer was the level of prices in relation to costs. The spiral of rising prices between 1760 and 1813 made it possible for many small men to carry on, men who would have failed before 1750 and who probably did fail in the difficult years after 1815, no doubt because of high costs, high rents per acre and the burden of the poor rates. It must also be remembered that many mortgaged their property to rebuild their farm-houses, find dowries for their daughters and place younger sons in a profession. The return to the gold standard in 1821, the reduction in the number of bank notes in circulation and the failure of many country banks caused lenders to call in their loans and sales of farms had often to follow.

'The effect on the cottager can be best described by saying that before enclosure the cottager was a labourer with land, after enclosure he was a labourer without land.' This was the Hammonds' definition of a cottager. There is some truth in this statement for the cottager had access to the waste and the common where he could keep some pigs, a cow, some geese, gather firewood and in the Pennine villages the right of turbary or cutting peat for household fuel. In the case of the squatters it provided a place for a house of a sort. When this prop was removed they faced increasing poverty. However, much depended on local conditions, the employment available, the situation of allotments, and the nature of the soil. Even before enclosure the majority of the cottagers had no common rights, for these rights were attached to open field holdings, or to certain cottages, whose owners or occupiers only could exercise them. Legal owners of common rights were always compensated by an allotment. The occupiers of common right cottages received nothing since they were merely tenants and not owners, since the latter always received the allotment. Unfortunately, the allotments for cottages were too small to be of practical use, being far smaller than the three acres needed for a cow and often at an inconvenient distance from the cottage, to which had to be added the high cost of fencing the allotment. Cottagers often sold their plots to neighbouring farmers, so peasant ownership at its lowest level declined. Customary rights were not legal rights, though commissioners sometimes recognised them and at other times ignored them. Arthur Young's remark that 'by nineteen enclosures out of twenty the poor are injured, in some grossly injured. . . . The poor in these parishes may say Parliament may be tender of property; all I know is I once had a cow and an Act of Parliament has taken it from me', can be challenged. Clapham states that the number of parishes adversely affected was far less than nineteen out of twenty for in many places the poor were left with cow pastures, and over England as a whole the majority of labourers were left with a garden or a potato patch (12). Allotments and gardens were scarce in those areas where enclosure was late.

What of the general statement that change meant a wholesale exodus of labour from the country to the town? This is by no means accurate. Where enclosure of the common and wastes expanded the area of cultivation there was a demand for labour in fencing, hedge planting, wall building, construction of new roads, and the building of new farmhouses and barns. Dr Joan Thirsk has pointed out that there was an acute shortage of labour in Lincolnshire when the fenlands were enclosed. Larger crops produced on these additional acres created the demand for labour for harvesting, winnowing and threshing. Connected with this expansion in agriculture was the expansion in the ancillary trades of flour-milling, barley-malting, hop-drying, leather-tanning, cheese-making, bacon-curing and in rural crafts, such as the wheelwright and the blacksmith. All these occupations expanded as the cultivated area extended and production increased. Professor Chambers has pointed out that in the Nottinghamshire villages, the population increase was only at a slightly lower rate than in the villages dominated by mining and textile industries (21). Poverty amongst the cottagers existed in those areas which had been enclosed by private agreement and not by an Act of Parliament and which had no waste land for allocation. Another factor was the operation of the Law of Settlement leading to closed parishes, where to keep down the poor rates newcomers were refused a permanent settlement. It is not easy to distinguish the influence of enclosure from among the other factors that affected unemployment, such as the expansion of the textile, mining, iron and steel industries, the migration of industry from East Anglia and Sussex to new areas, and the recurring periods of boom and slump in trade. In addition there was the effect of the weather on harvests and the growth of permanent pasture at the expense of arable, all of which played a part. Overall it was the population explosion that caused poverty. The labour supply increased faster than agriculture could absorb it, so that a pool of landless and unemployed persons appeared in both open and enclosed villages. In the rural counties of the south the unemployment situation among farm labourers became so bad by 1830 that riots, rick burning, and the activities of 'Captain Swing' created serious disturbances which had to be put down by force (82). The ending of open access to the commons and wastes led to the composition of scurrilous verses [doc. 56].

There had always been a good deal of rural poverty, and Dr Joan Thirsk has emphasised this factor in her study of the Hearth Tax returns in Leicestershire for 1670 (52). The same pattern can be observed in the West Riding villages, where some domestic industry was usually available to ease the situation. The greater part of the unemployed came from the growing numbers of younger sons with no land to inherit, so that the growth of population was the main factor in the increase in landless and partially workless labour force in the counties.

The growing demand of townsfolk for agricultural produce is described by Defoe in his *Tour of England and Wales* (2). He described the production of turkeys and geese in East Anglia for the London market. He noted the contribution made by various counties in the production of meat, corn, butter and cheese to satisfy the demands of the town markets in their areas. In the Midlands, Lancashire and Yorkshire the industrial areas obtained their produce from the immediate rural areas and the neighbouring counties. Milk could not be transported more than ten miles, so dairies were situated in the towns and the cows were milked in the streets. No notice was taken of the lack of hygiene in this system [doc. 57].

The growth of the textile, iron, steel, coal, hardware and pottery industries which were concentrated in the West Riding, Lancashire, the Black Country, the north Midlands and the north-east coast affected the development of northern agriculture. Liverpool, Manchester, Hull, Leeds and Newcastle advanced in importance and population. Arthur Young was concerned about the neglect of agriculture by the West Riding towns and feared that, unless something could be done to change this attitude, starvation faced the region in time (14). The Pennines, Cheviots and Westmorland fells supported sheep, which supplied wool to the cloth manufacturers of the West Riding and east Lancashire, as well as mutton to feed the miners of Durham and Yorkshire, the cutlers of Sheffield and the spinners and weavers of Huddersfield, Leeds and Bradford. Scotch and Irish cattle were brought to fatten on the grasslands to satisfy the growing demand for meat in the local markets. Robert Brown, reporting in 1799, stated that the West Riding was unable to supply itself with food as the population increased [doc. 18]. Potatoes were grown as a commercial crop in areas close to urban centres, while quantities of mutton and beef were demanded in the towns. Horses were in great demand for working the horse gins at coal mines, hauling wagons, and as pack animals over the Pennines. In 1784 a train of twelve pack horses went from Richmond to Lancaster to transport the court records of the Archdeaconry of Richmond from Lancaster. Townsfolk lost their open spaces, and their racecourses, except at Doncaster and a few other towns, disappeared. The sites were utilised for the building of masses of congested, back to back houses and factories. Amenities for open air enjoyment ceased, and although the working man's wages were higher and his purchasing power greater, his leisure activities — apart from drinking in public houses — were seriously reduced (16).

6 Agriculture and Industry

It is necessary at this point to examine briefly the relationship between agriculture and industry. English landowners have always accepted that land is an agent of production and not merely a source of social and political prestige. Through their system of estate management and the employment of able and progressive agents and stewards, whose part in agricultural development has yet to be studied, they created a profitable industry [doc. 59]. The correspondence of such persons as Earl Fitzwilliam, the Marquis of Rockingham, Lord Halifax and Lord Harewood with their stewards reveals the depth of interest and concern about their estates [doc. 60].

Through their work the landowners got the best of the political world and social standing, which was conferred by land ownership and also of the commercial world which good tenants could open to them. There developed a farming system that could respond to market demands and which, under the impact of population increase, acquired a flexibility that made English farmers world famous [doc. 61].

Landowners with political power were ideal agents for such development schemes as required Acts of Parliament. As a result they became interested in river navigation, turnpike trusts, canals and later railways, which improved the transport of grain, cattle, lime, manure and bricks for erecting buildings both in towns and in the countryside. The contribution to this effort of the small farmer who did the carting of iron, coal and stone during slack periods on the farm, must not be ignored [doc. 62]. Industrial growth was dependent on the transfer of scarce human and material resources from farming to industry. Coal transport was a major problem and landowners in the north, who were also frequently coal owners, were interested in improving transport. Hence it is not surprising to find that of the total amount of money subscribed in canal stock between 1758 and 1802 a very large proportion came from the pockets of the landed prorietors. One of these was John Spencer of Cannon Hall, Barnsley, a keen estate improver and at the same time involved in iron, coal, wool and lead industries in both Yorkshire and Derbyshire. Another was Robert Clarke of Silkstone, who ranks alongside Earl Fitzwilliam, Lord Halifax and Thomas

Beaumont of Bretton Hall, whose families were destined to play an important part in the development of the railway system in England. At the same time the activities of the Lowthers in Cumberland and Lords Durham and Londonderry in Northumberland are reminders of the aristocratic origin of industrial enterprise north of the Trent.

Professor A. H. John has argued that low wages before 1750 raised real wages later and general standards of living rose which in turn stimulated the home market and formed the prelude to the Industrial Revolution (71). This meant better houses, clothes, household goods, meat and dairy produce, as well as wheat bread for the table. Professor Chambers states that 'a good harvest called for a huge army of harvesters, men, women and children to cut, stook, bind, load and carry sheaves to the stackyard'. In good years wages rose and prices fell, with non-agricultural products becoming more expensive as labourers spent their surplus earnings. Tradesmen who used agricultural produce, merchants, millers, bakers, starch-makers, stationers, bookbinders, linen-printers, trunk makers and paper-hangers, all benefited from cheap flour. Cheap barley helped the distillers, maltsters, brewers, and innkeepers. An abundance of animals benefited shoemakers, harness makers, soap boilers, chandlers, cutlers and gluemakers. Spinners and weavers had abundant wool. Workers sold their labour to manufacturers at higher rates and all had more to spend. Farmers and landowners grumbled when prices fell and wages rose but the reverse happened when harvests were poor and incomes fell with rising unemployment (17, 21). The landowner of 1850 lived better than his ancestor in 1750 and the same is also true of the labourer in the Midlands and the north, but not in the low wages area of the south and east.

The end of the Agrarian Revolution was the end of an era and the repeal of the Corn Laws in 1846 heralded twenty-nine years of development in agriculture on a large scale. The thirty years following the repeal of the Corn Laws was a period of expanding markets, the growth of industrial towns and improved transport. It was a period in which farming made rapid progress. In 1838 the Royal Agricultural Society had been established to give farmers advice on improved methods in farming practice. Seven years later the Cirencester Agricultural College was founded to give a better training to farmers. Justus von Liebig experimented with sulphuric acid on bones in order to produce phosphates which would serve as an artificial fertiliser and a factory was established at Deptford in 1842 to produce these fertilisers. In 1843 Sir John Lawes had founded the agricultural research station at Rothamsted. Drainage remained a problem until John Fowler of Leeds developed his steam mole plough for draining which, with the cheap land tiles produced by Thomas Scragg's machine, lowered the cost of draining land considerably. This improved system of drainage helped farmers on heavy clay soils to cut their costs, speed up their operations and introduce root crops on a larger scale. After 1846 the government came to the aid of farmers by granting them loans at a low interest rate of 3½ per cent to drain their land. Not only was the soil improved by draining and by the use of artificial fertiliser, but also by imports of Peruvian guano, Chilean nitrates and basic slag. It was also possible to introduce improved cattle foods by the use of mangel wurzels, kohlrabi, maize and cattle cake made from cotton seeds and linseed oil, all of which supplemented the use of grass and hay. Farm building was also easier and cheaper after the tax on bricks was removed in 1850.

The expansion of the railway system in England had opened up remote counties to the population of the industrial areas and made it possible to transport farm produce hundreds of miles at a cost no greater than it had been to transport the same goods some thirty miles under the old system. As the railways became more efficient the droving trade declined, for rail transport meant that cattle, sheep and fowls could arrive fresh at the market without loss of weight.

The nineteenth century saw farming become a machine using industry. Though reaping by machine was not widely practised until after 1850, a Northumberland millwright, John Common, had in 1812 designed a reaping machine which McCormick later developed for use on the American prairies. The reaper designed by the Rev. Patrick Bell in 1826 began to come into use in 1853, having been improved and marketed as the Beverley Reaper. Andrew Meickle's threshing machine could be driven by steam, water, horse or hand power, and it shortened the time, as well as the cost, of threshing barley, wheat and oats. The introduction of this machine in the south in the late 'twenties deprived many labourers of their winter threshing and contributed to the troubles of 1830. Chaff cutters and turnip choppers were at first simple hand operated machines but were later harnessed to steam power.

From 1850 onwards farmers were beginning to use extensively a wider range of efficient machinery, which was widely advertised. The development of the traction engine by John Fowler of Leeds and Aveling and Porter of Rochester meant that the threshing machines could be moved from place to place as required. Machines were cheap at prices ranging from £10 to £20 each. Horse-drawn drills and cultivators extended the area of cultivation. In 1859 it was possible to buy a fourteen horsepower Fowler steam plough which could do the work formerly undertaken by seven four-ox plough teams. Since these machines meant that larger fields were necessary, many of the small fields allotted and fenced by an enclosure award had their hedges pulled up and were made into larger fields. By 1870 there were more than 40,000 McCormick reapers in use in England. All these improvements were made possible by war in Europe and America, high transport charges and technical problems, which prevented both America and Russia from exploiting their fertile soils. By 1875 the picture had changed. Railways and machine harvesting had opened up the prairies of the Middle West, while the elevator system of loading ships, coupled with improved steamship design, reduced the cost of transport of wheat from Chicago to Liverpool by 75 per cent. Hence after 1875 British farming had to face keen foreign competition and the latter years of the century saw farming once again change its pattern in England. The years of depression after 1875 saw new lines of development.

The Agrarian Revolution is not the counterpart in rural areas of the Industrial Revolution. It was a long process of change in farming methods, in the use of land and the expansion of the cultivated area to give bigger yields. The development had taken place over many centuries as rising populations, or the reverse, influenced change in agricultural practices. The breakdown of arable into grassland in the sixteenth century had met the demand for wool. This was followed in the seventeenth century by the practice of laying arable land down as temporary grassland with the introduction of clover and grasses. In the eighteenth century new methods developed which saw the emergence of the enclosed, compact, economic farm unit which was exploited to the full in the middle years of the nineteenth century. So often the enclosure Act and award is but the logical conclusion to a piecemeal long-term development stretching back in some cases to 1285, if not earlier, as geographical factors and population pressures brought about change.

Assarting in Woolley and Skelmanthorpe

Documents 1 and 2 are examples of grants of land common to many parts of the country, especially where family papers have survived. They give detailed information about the situation and are clearly distinguished from the crofts.

1284; 4 November: Grant by Pauline de Skelmert' to Adam de Holande of Maude, widow of Skelmert', and all her following and her house of three perches belonging to that house and the assart Sygardia held which lay between Hyresdene and Torp and between the assart held by Fyrgus for his homage and service. . . .

Yorkshire Deeds, vol. x, Yorkshire Archaeological Society Records Series.

1392; 29 August: Grant in tail by John Staynton of Woolley and Elizabeth his wife to their son John of a built messuage in Wolaymorehouses except an assart called Northroyd which the grantors lately acquired of Alice daughter of . . . Adam Sprigonell . . . and also of three assarts lying separately in Woolaymorehouses called Spytilcroft, Sarcroft and Rydinges. . . .

Ibid.

Permission to enclose land for a park

Medieval monarchs were always willing to raise money by granting charters to loyal subjects to change the use of land.

John, by the grace of God, King of England, Lord of Ireland, Duke of Normandy and Acquitaine, Count of Anjou . . . to all his faithfull subjects greeting. Know that we have granted and by our full charter confirmed to Robert le Vavasour that he may have free warren throughout the whole of his land in Werverdale, of all beasts and that he may make a park there if he so wishes. Wherefore our will and command to the aforesaid Robert and his heirs after him is that he may have and hold that warren and park if it has been made in the aforesaid land as decreed, sound, in peace, free and quiet and entire . . . Datum per manum . . . 13 Day of March in the fifth year of our reign [13 March 1204].

Vavasour Papers, Leeds City Archives.

Vavasour Papers, Leeds City Archives.

document 4

Effect of the Black Death on Selby Abbey

This extract from a letter sent by the Abbot of Selby shows some of the problems that faced a monastery after the Black Death.

To his most excellent Prince and Lord . . . the Lord Edward . . . King of England . . . his most humble chaplain Geoffrey, Abbot of the monastery of Selby. . . . Since we are occupied beyond our strength in supporting the charges incumbent on our monastery as well because our discreeter and stronger brethren . . . have gone the way of all flesh through the pestilence, as because our house both in decay of rents and in lack of corn and other victuals is suffering undue disaster . . . we are unable to be present in the instant Parliament to be held on the Octave of the Purification of the Blessed Virgin Mary next coming.

Duchy of Lancaster Miscellaneous Books, 8f. 57d, P.R.O.

document 5

The manor under duress: extracts from the Court Roll of the manor of Thorner in the year of the Black Death

This roll is in many places faded and obscure; many of the entries are heavily interlineated, altered and cancelled. The membranes have been sewn together with no vestige of order apparent. The extracts show the confusion that exists.

THE COURT OF THOMAS DE METHAM HELD AT THORNER●.. DAY [*obscure*] AFTER THE FEAST OF ST WILFRID IN THE 23RD YEAR OF THE REIGN OF KING EDWARD THE THIRD AFTER THE CONQUEST.

William le Smale comes into court and makes fealty to the lord and acknowledges his tenure from the lord of one messuage and 26 acres of land; he holds another messuage but does not know by what service.

Ellen, sister of one Margaret de Merlay, comes into court and makes fealty to the lord and acknowledges her tenure from the lord of a house, 9½ acres of land and a rood of land, by service of 12d. per year suit of court twice a year, reaping for a day and a half in autumn, ploughing with an iron tyred cart for 2 days in the year ... helping to make the Milnedame just as the other tenants do, and outside service.

William de Husum comes into court and makes fealty ... for 4 acres and 1 rood by outside service and suit of court from 3 weeks to 3 weeks ... and it is taken back into the lord's hands because he is under age.

It is ordered to take into the lord's hand the tenement which was Robert del Hille's, who died without heir; later John del Hille came and claimed tenure ... being as it were brother and heir.

It is ordered to take into the lord's hand the tenement of Alice and John the children of Thomas Smith by reason of their minor age. It is ordered to take the tenement of Robert Prior by reason of the minor age of Thomas son of the said Robert.

Earl of Mexborough muniments, Leeds City Archives.

document 6

Leasing the demesne, 1326

By the fourteenth century some lords were leasing off parts of their demesne lands since it was easier to cultivate them by hired labour and only boon work was retained.

Lease from Henry Vavasour to Gilbert Laitlove of Stutton, Matilda his wife and Walter their son of a messuage, two bovates of land and four acres of arable land from his demesne in the vill and district of Cottesford rendering 12s. 10d. also one mark to the altar of St Mary the Virgin in Heselwode chapel. And they shall do three boon works per year with their plough, four harrows and two reaping women for one day in autumn; and one meal per day for each and every boon work. They shall make suit at Henry's court at Cobblesthorpe twice a year. Witnesses; John de Lascy, John of Acclesthorpe, Robert of Saxton and Robert clerk of Stutton.

Vavasour Papers, Leeds City Archives.

document 7

Increase in prices, 1550

This account is an attempt to discover the cause of the increase in the cost of living and the prices of goods.

With the alteration of the coin began this dearth; and as the coin appeared, so rose the price of things withal. And this to be true the few pieces of old coin yet remaining testifieth; for ye shall have for any of the said coin, as much of any ware either inward or outward as much as ever was wont to be had for the same. . . . And because this riseth not together at all men's hands, therefore some hath great loss and some other great gains thereby, and that makes such a general grudge for the thing. . . . I think this alteration of the coin to be the first original cause that strangers first sold their wares dearer to us; and that makes all farmers and tenants that reareth any commodity again to sell the same dearer; the dearth thereof makes all gentlemen raise their rents . . . and consequently to enclose more grounds.

A Discourse of the Common Weal of this Realm of England, 1581 (ed. E. Lamond, 1893, p. 104).

Enclosure defined

John Hales, a government official, attempted to reverse the enclosure movement. This extract is from an address he made during his enquiry into enclosures in the Midlands in 1548. The commissions of enquiry failed to overcome the resistance of the powerful gentry who wanted enclosure.

To declare unto you what is meant by the word enclosure. It is not taken where a man does enclose and hedge his own proper ground where no man has commons. For such enclosure is very beneficial to the commonwealth: it is a cause of great increase of wood. But it is meant thereby when any man has taken away and enclosed other mens commons, or has pulled down houses of husbandry and converted the lands from tillage to pasture. This is the meaning of the word. . . . To defeat these statutes as we be informed, some have not pulled down their houses but maintain them; howbeit, no person dwells therein, or if there be it is but a shepherd or a milkmaid; and convert the lands from tillage to pasture. And some sow about one hundred acres of ground, or more or less, make a deep furrow and sow that and the rest they till not but pasture with their sheep. And some take the lands from their houses and occupy them in husbandry, but let the houses out to beggars and old poor people. Some to disguise the multitude of their sheep father them on their children, kinsfolk and servants. All which be but only crafts and subtleties to defraud the laws, such as no good man will use but rather abhor. . . .

Besides it is not unlike but that these great fines for lands and improvements of rents shall abate, and all things wax better cheap − 20 and 30 eggs for a penny and the rest after the rate as has been in times past. . . .

J. Strype, *Ecclesiastical Memorials*, 1721, vol. ii, Doc. Q.

Inquiry into depopulation, 1570

The ignorance of the jurors at the inquest into this village indicates that depopulation had taken place and it had been imparked by the middle of the fifteenth century.

East Lilling it is called, and retaineth the name of East Lilling township, though at this day there do remain only one house wherein the said Mistress Hall now dwelleth, a competent house for a gentleman. . . .

But by tradition and by apparent ancient buildings and wayes for horse and cart visibly discerned leading unto the place where the two stood, within Sheriff Hutton Park, it hath been a hamlet of some capacity, though now utterly demolished and the place where it stood dismembered from the present territories of East Lilling and is now part of and impaled to the park of Sheriff Hutton, how long since doth not appear.

P.R.O. E179/211/30m. 18.

document 10

Thomas Cromwell to Henry VIII on sheep farming

The concern of the government is shown in this extract from a letter sent by Cromwell to Henry lest sheep replace arable farming.

. . . It may please your most royal Majesty to know that yesterday there passed your Commons a bill that no person within this your realm shall hereafter keep and nourish above the number of 2000 sheep and also that the eighth part of every man's land, being a farmer shall for ever hereafter be put in tillage yearly.

P.R.O., Record Office Calendar, vii, m. 73, 1535.

document 11

Tithe dispute at Danby Wiske, 1574

The excuse for commuting the tithes was stated to be the rising of the northern earls in 1569. Part of the evidence of two witnesses is reproduced here.

Adam Tenant stated that the commutation of corn and hay tithe was the direct result of the rising of the Northern Earls in 1569 and continued, '. . . that by reason of the troble in that yere verre fewe of the parishioners of Danby Wiske did pay their tithes in kind for that corne and haie was destroyed and wasted greatlye in the parish the same time of the Rebellion and therefore the parson did compound with all or most of the parishioners to take in money of everie one much lesse than the tithes were worth. . . .'

John Kirby the schoolmaster of Thirsk contended that this was not so and said, '. . . there was no corne or haie destroyed in the parishes because the Rebellion was not begun until Martinmas following the harvest . . . he dwelling in Bedale for thre yeres teaching a Scole ther at the same time. . . .'

Tithe Cause Papers, R.VII G, No. 1689, Borthwick Institute.

document 12
Enclosure at Thurgoland, 1577

The tenants are afraid that if enclosure continues all pasture will disappear so an order is made for the enclosures to be destroyed but this was largely ineffective.

Item . . . that John Cudworthe of Eastfield shall lay out againe . . . one enclosed parcell of ground nowe in his occupaycion which hath bene enclosed of the lords waste before the feast of Seynt John Baptiste next coming commonly called Mydsomer Day . . . sub pena 6/8.
Item that Thomas Cudworth of Coates shall cast out and suffer to lye open & unenclosed certain parcells of ground which he hath enclosed from the lords waste conteyninge by estimacion A Rood . . . sub pena 20d.

Thurgoland Manor Rolls, 1577, Spencer Stanhope MS 60230, Sheffield City Library.

document 13
Enclosure at Skelmanthorpe, 1636-40

Here in this extract from the Court Roll of 8 June 1636 it can be observed that the decision to enclosure has been privately agreed and instructions are issued as to procedure.

It is ordered by this Courte that the tenants and occupiers of the fowre Common townefields lately devyded shall on this side and before the feast daie of the Annunciacion of our Lady St Mary the blessed Virgin next . . . make sufficient their several partes of the Fences and hedges thereunto belonginge and the same being soe made shall at all tymes hereafter soe meyntayne and keep the same.

Item that this Courte doth order that the tenants or occupiers of the 3 Common townefields not devyded shall on this side and before the 25 daie of March which shal be in that yeare of our Lord God 1637 devyde the said 3 townefields, And everie tenant before the 25th daie of March . . . 1640 make his several parte of such fence . . . as shalbe allotted to him to make or meyntayne. . . . And afterwards everie of them shall keep their severall partes of the same . . . in good repair.

Skelmanthorpe Manor Court Rolls, 1636-40, Spencer Stanhope MS 64726, Sheffield City Library.

document 14

Advice to farmers, 1523

The boom in farming in the sixteenth century demanded efficient management and several writers, of whom Anthony Fitzherbert was one, devoted themselves to this task of spreading the required knowledge.

An husband cannot well thrive by his corn without he has other cattle, nor by his cattle without corn: for else he shall be a buyer, borrower or beggar; and . . . sheep in my opinion is the most profitableest cattle that a man can have. . . .

A shepherd should not go without his dog, his sheep hook, a pair of shears, and his tar box either with him or ready at his sheep fold. And he must teach his dog to bark when he would have him, to run when he would have him, and to leave running when he would have him or else he is not a cunning shepherd. The dog must learn it when his is a whelp, or else it will not be; for it is hard to make an old dog to stoop. . . .

Now thou husband that hast both horses and mares, beasts and sheep, it were necessary also that thou have both swine and bees. For it is an old saying; he that has both sheep, swine and bees, sleep he, wake he, he may thrive. And that saying is because they be those things that most profit rises (from) in shortest space with the least cost. . . .

Sir Anthony Fitzherbert, *The Book of Husbandry*, 1523.

document 15
Defoe's journey from Halifax

Daniel Defoe described his experiences on a tour through England in 1727. In this extract he describes the difficulties of travelling from Halifax to Leeds.

We quitted Halifax not without astonishment at its situation being so surrounded with hills and those so high as (except the entrance by the west) makes coming and going exceeding troublesome and indeed for carriages hardly practicable and particularly the hill up which they go out of the town east-ward towards Leeds and which the country people call Halifax Bank, is so steep, so rugged and sometimes so slip-pery that to a town of so much business as this 'tis exceeding troublesome and dangerous.

Daniel Defoe (2).

document 16
Arthur Young's journey from Preston to Wigan

This is a classic description of the state of the main roads in Lancashire when Young was making his tour in 1778.

To look over a map and perceive that it is a principal one not only to some towns, but even to whole counties, one would naturally conclude it to be at least decent; but let me most seriously caution all travellers . . . to avoid it as they would the devil for a thousand to one but they break their necks or their limbs by overthrows or breaking down. They will meet here with ruts . . . four feet deep floating with mud only from a wet summer; what, therefore must it be after winter? The only mending it receives is the tumbling in of some loose stones which serve no other purpose but jolting a carriage in the most intolerable manner.

Arthur Young (15).

document 17

Returns on land use in Yorkshire parishes, 1795

These returns were the result of a survey and enquiry into the state of agriculture and the production of crops in time of war. Three different areas have been selected.

EMLEY, population 1,642; 1,275 acres grass, 416 wheat, 597 oats, 96 barley, 22 potatoes, 96 turnips, 96 beans and peas.
BURTON LEONARD, population 270; 600 acres grass, 800 arable, 260 wheat, 290 barley or oats, 250 turnips and potatoes.
HATFIELD, population 2,000; 3,858 acres grass, 1,180 wheat, 1,145 oats, 75 beans, 592 barley, 67 potatoes, 300 turnips.
Total acreage of the three parishes is 13,650 acres.

R. Brown, *General View of Agriculture in the West Riding of Yorkshire 1795* (edition 1799).

document 18

Prices of provisions in Yorkshire 1795

A report on the prices of provisions and the state of agriculture in Yorkshire was compiled in 1795. The following is an extract from the report on the West Riding.

As the West Riding, from the extent of population, is unable to supply itself with provisions, the prices are full as high as in any part of the island. . . . At Wakefield market in July 1795, wheat was sold at the enormous price of £9 per quarter; and it may be remarked, that during such critical periods, the county which depends upon foreign supplies, must comparatively pay much higher prices for the articles which cannot be furnished within its own bounds, than what they do in ordinary seasons; and that prices must necessarily advance to a far higher rate than is usual in those counties, where the articles are produced. The scarcity is felt in a serious way, and it requires great exertions to provide a supply. . . .

R. Brown, *General View of Agriculture in the West Riding of Yorkshire 1795.*

Description of convertible farming
document 19

William Marshall made reports on farming practices in various parts of the country. The following is part of his report on the Midlands.

The outlines of management consist in keeping the land in grass and corn alternately . . . and in applying the grass to the breeding of heifers for the dairy, to dairying and to the grazing of barren and aged cows; with a mixture of ewes and lambs for the butcher. . . . The land having lain six or seven years in a state of SWARD provincially 'TURF' — it is broken up by a single plowing for OATS, the oats stubble plowed two or three times for WHEAT: and the wheat stubble winter fallowed for BARLEY and GRASS SEEDS: letting the land lie another period of six or seven years in HERBAGE: and then again breaking it up, for the same singular SUCCESSION OF ARABLE CROPS. . . .

William Marshall (6).

Arthur Young's visit to Wentworth
document 20

Arthur Young was an admirer of the Marquis of Rockingham as a model farmer and estate owner.

Being well convinced that argument and persuasion would have little effect with the John Trot geniuses of farming, he (the Marquis) determined to set the example of good husbandry as the only probable means of being successful. In the pursuit of this end his Lordship's conduct was judicious and spirited. He has upward of 2,000 acres in his hands and began their improvement with draining such as were wet, rightly considering his part of husbandry as the sine qua non of all other. . . .

Arthur Young (14), i, 308-9.

document 21

Report on crop production at Wentworth, 1753

The Wentworth Estates kept detailed records of production from each farm. Hence these records are valuable for a study of period farming.

Tankersley; 10 acres of turnips.
    John Sampson grew   5 acres of pease, 3 acres of line, 10 acres of clover.
    Joseph Trippet grew   1 acre of pease and beans and 3 acres of clover.

Wentworth Woodhouse Estate accounts, A223, 1753.

document 22

Rockingham Memorandum Book, 1753

In this book the Marquis of Rockingham recorded details of all his experiments in crop rotation, drainage and manuring, with results.

The Hankerchief Piece was manured in October 1753 on the 3 Acres next the Roads, 2 Load of Nottingley Lime per Acre on the 2 Acres above them, two load of Rotherham lime mix'd with 2 load of Tan Bark on the 2 Uppermost acres, 2 load of Hooton lime mix'd with 2 loads of Tan Bark.
    Compost according to the proportion mentioned by Houghton No. 14 for One Acre.

half a load of Hooton Lime		4s	0
2^2 of a load of Pidgeons' dung		5s	0
abt. 5 stone of Salt Petre	5	5s	0

£5 ; 14 ; 0

Compost to be tried per acre.

half a load of Hooton Lime	4s	0
2^2 of a load of Pidgeons' dung	5s	0
3 Bushels of Common Salt	16s	0

£1 ; 5 ; 0

First Paddock on the Newbiggin Side. Turf to be pared and burnt, well drained and Sown with Turneps this Winter & in Spring Sown with Sainte foin or Rye Grass.

Few Acres of upper Park of Tankersley Park to be tryed on Same Manner. Lime in Clay Grounds will produce as good Turneps as dung.

Tillage

Tull says St Foin should be sown early in Spring drilled in Rows of 8 inches as under & abt. 1 Bushel per acre. The Seed should not be above half an inch under ground.

Propose to try of the Land in Tankersley Park which shall be pared & burnt five acres according to this method & also five acres after the following Method which I learnt at Marshfield had been tried & succeeded.

Pare & Burn in the Summer early. Sow it with Turnips & in the Spring late, sow 4 Bushel of St Foin Seed per acre.

Main Carrier Drain to run Cross Temple Close between Troules Wood & Shire Oaks & thro' Shire Oaks. Several to be cut in the first Paddock on the Newbiggin Side filled with Stones like those in the Horse Closes only Deeper in order to plough over & some of them 8 inches broad at Bottom.

Marquis of Rockingham's Memorandum Book, R.183.

Letter from M. Banks to John Eccleston

This letter shows the keen interest taken by George III in farming and in this case offering a Merino Ram to improve the breed of sheep in Lancashire.

Soho Square
Aug. 15 1792

My Dear Sir,

Having represented to the King that our Friend Mr Eccleston is a great promoter of improvement in every kind of Agriculture I was honard on Saturday last with his Royal Command to chuse from his Spanish Flock a Ram of the true Merino Breed & deliver it to his order wherever he is Pleezd to send to Windsor for it.

His Command you may be asured I cheerfully Obeyd & I have chose for him a Ram imported from Estremedura which was used in the Kings Flock last year, he is 5 years old but I could not get an original one younger nor do I believe (but that must not be pronounced) that a better sheep of its Kind is to be Found in his Majesties Posession.

Will you be so good as to acquaint our Friend of the Circumstance & tell him that the Sheep will be delivered to his order by Mr Robinson of Windsor who is to be heard of at the Queens Lodge or the Maestrick Gordon & that he is mark'd on the ham with the Number 1.

believe me dear sir
Most Faithfully Yours
M Banks.

Scarisbrick Hall Papers, DDSc 9/42, Lancashire Record Office.

A colliery lease at Bilham Grange, 1659

It was important that harvest should be completed as quickly as possible before the weather broke. Hence written into leases, as in the following, were conditions about the share of the harvest work to be undertaken by the colliers.

Articles of bargain and demise made, concluded and agreed upon by and between Richard Allott of Holmfirth in the parish of Kirkburton in the County of Yorke, gentleman, of the one part and John Punder, Christopher Wildsmith and John Woodhead all of Clayton in the Parish of High Hoyland in the Said County, Colliers of the other part the eleventh day of April in the year of our Lord 1659. . . . And when the next corn harvest comes will and shall sheare and cutt downe all the corne of the said Richard Allott in Clayton aforesaid he the said Richard Allott paying or allowing out of the said Rent unto them one pound nineteen shillings and eight pence for the said harvest work. And shall and will also in haytime and harvest during all the said time (as often as neede shall require) he the said Richard Allott shall call or send unto them to labour and work with him in any manner of worke to or for the inninge of his hay or corn at Bilham Grange aforesaid and he the said Richard Allott paying twelve pence per diem unto the said Christopher Wildsmith and unto the said John Punder and John Woodhead eight pence per diem unto eyther of them and they shall not work with any other work during the said time without the said Richard Allott's permission. . . .

Bretton Hall Archives, DD70/60, Yorkshire Archaeological Society.

document 25

Presentment to the Archdeacon, 1723

Annually at Easter the churchwardens of each parish had to present to the Archdeacon a list of all the faults and weaknesses of both clergy and laity as well as details of the conduct of the priest.

Contra the Vicar that he did not say Evening Prayer on the last Sunday in September in the afternoon but took the congregation to get in his Barley the weather being fine.

Archdeacon of York's Court Book, 1723, Borthwick Institute of Historical Research.

Elmhirst farm accounts, 1825-26

This selection from the ledger of the farm accounts is to show the costly and slow method of harvesting and threshing. Note the low rate of wages.

1825

July 15.	Mary Parkin 16.9d for 11¾ days wages. 10 days at hay in Clay, Round and Green Closes.
	Mary Fisher 12.0d for 8 days wages making hay.
Sept. 3	5.0d for Ale for harvest when leading corn.
Dec. 14	Rec'd of John Rolling £26; 18; 4d for 20 loads of wheat threshed by Dewsnap at £1; 6; 11d a load.

1826

August 10	paid Thomas Parkin 16.6d for 3 loads of Manure for Sweed.
	paid John Jackson 11.6d for an acre of Wheat Shearing in Far Ox Close.
	paid Jackman & I Douglas & Dewsnap £6; 18; 0d for 34 acres of Grass mowing in Great Close and Paddock.
	paid James Douglas 15.9d for 5 acres oats mowing at 3d an acre.
August 12	paid Thomas Dewsnap & Swift £2; 2; 0d for pulling 6 acres of Beans at 7.0d per acre.
August 14	paid Ben Huddleston for Shearing Wheat 17.3d at 11.6d an acre.
	paid John Jackman 13.6d for 6 acres of Beans pulling at 2.3d an acre.
August 21	paid Thomas Winter 5.9d for one half acre of wheat shearing at 11.6d an acre.

Elmhirst Archives, EM.1187, Sheffield City Library.

Enclosure at Mexborough

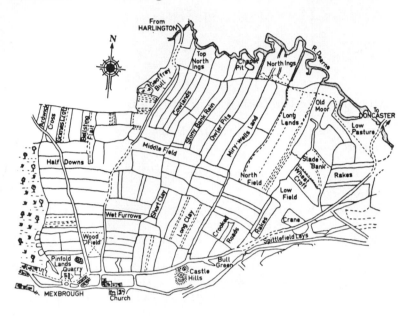

Encroachment on the lands of Hoyland Swaine, 1614

This is part of the evidence from a case heard in Star Chamber which was promoted by Sir Francis Wortley, Lord of the Manor of Hoyland Swaine against the villagers of Cawthorne for moving the boundary stones, between the two villages.

. . . that the lordes inheritors and inhabitants of Cawthorne aforesaid having longe tyme before the memory of man enclosed the greatest part of the landes in Cawthorne being low and good ground of itself a greate parte of ground in Hoyland Swaine being barren and lyinge highe and wanting good means to enclose and yet doe lye open to the Commons of Hoyland Swaine. Thomas Greene of Cawthorne, John Gawthorpe, Edmund Wainwright and Anne his wife, Thomas Ellis and divers others have removed the meerstones and

73

broken the same in pieces and carried them away to intrude on the Common of Hoyland Swaine.

Wharncliffe Papers, Wh.M.21, Sheffield City Libraries.

Letter from William Earl of Strafford about enclosure

This letter asks for advice about the leasing of coal pits and the enclosure at Tankersley.

For Sir Thomas Wentworth of Bretton these.

Good Cosen,
    So soone as you can conveniently, I desire you will doe me the kindnesse to goe to Woodhouse. . . . Also when you are there I desire you will doe me the favour to make me as good a bargaine as you can in letting my cole pitts; and that you will acquaint me with your opinion about some petition upon my father's inclosing of Tankersley. I have appointed Bower to send you them and before I give them an answer I am desirous to heare your advice.
                                    Knowsley 29 October 1663

A. K. Clayton (**22**).

Private Enclosure at Mexborough, 1737

These two extracts are taken from the depositions of the witnesses in a case brought into Chancery by William Sylvester against Charles Savile concerning a disputed enclosure. The first passage gives details of farming practice and the second the conditions laid down when land was enclosed at Mexborough by agreement.

George Cooper of Waddingham aged 60 says he was born in Mexborough and knows the fields called Wood Field, Middle Field & Lowfield and the parcels of ground in them called Wheat Croft, Crane, Crooked Roods and Rakes. One of the fields was always fallow every year & the other two are plowed and sown with corn & the same course of Husbandry is used with respect to the fallowing and sowing of Wheat Croft, Crane, Crooked Roods and Rakes. The rest is used with respect to fallowing & sowing of open fields.

Mary Savile aged 64 of Swinton widow says she knows the Bell String Flatt and heard her late husband say that the piece of ground was inclosed by agreement with the Inhabitants of Mexborough that Mr Savile would find Bell Ropes for the Church when the said Bells wanted Ropes and Mr Savile since he became owner found and provided Bell Ropes when there was occasion. . . .

Mexborough Inclosure Dispute, 1737, Spencer Stanhope MS 60215.

document 31

Rent adjustments after enclosure, 1751

When land was enclosed privately from the waste it was the custom for the rent to be adjusted accordingly.

1751.   Mr Jno Hill for a parcell of land inclosed from the Ranch Croft; 8s 0d to be added to the rent.

1757.   Thomas Hill 5s 0d rent for 0 acres 2 roods 16 perches of old enclosure from the waste and 8s 0d for three roods per consetum 1 Oct. By Thomas Hobson for walling and getting stones for a New Enclosure on Hoyland Common and a Bridge building £25; 12s; 8d.

Wentworth Woodhouse Muniments, WWM A222.

document 32

Private enclosure in Brampton and Greaseborough

A report drawn up on the state of the Wentworth estates of Earl Fitzwilliam in 1750 gives details of lands enclosed privately and the terms.

BRAMPTON N.B. By deed dated 19 March 1714 (No. 24) The Honble Thomas Wentworth and the Freeholders in Brampton agree to enclose Hoober Common whereby 60 acres is allotted to Mr Wentworth in lieu of a Coney Warren and 187 acres as his proportionate share in respect of his Estate in the Manor of Brampton.

GREASEBOROUGH By Ind're date 20 November 1717 (No. 14) The Freeholders in Greaseboro Release their Right in Greaseboro Common to Sir Thomas Wentworth who conveyed back to each Freeholder their share as set out and proportioned by Messrs Wharam & Hirst. And by Deed dated 16 August 1728 and a Fine levied in Trinity Term 1728 of the said Common Lands 261 acres part thereof enclosed and taken in for a Park is settled on Sir Thomas Wentworth as his share of the sd. Common & what he purchased from the Freeholders.

An Account of . . . Estates in Wentworth, Muniment Room, Wentworth Woodhouse (A. K. Clayton, 22).

document 33

Letter rejecting request to make boundary fences

24 June 1714 Darley Hall
Dear Sir,
I have received Mr Wortley' reply to the Letter you wrote respecting the Boundary Fences between Thurgoland and Wortley. The Inclosure of Wastes in Wortley was under a private Agreement made amongst the parties interested therein & Mr Wortley conceives that the Owners of Estates in Thurgoland have no Authority to call upon those in Wortley to perform anything contained in their Agreement or the Award made in pursuance of it, therefore the Wortley Proprietors decline Compliance with the request of making the Fences alluded to.

Your obdt Servant
Charles Bowns

Thurgoland Enclosure, Wh.M.61-4, Sheffield City Library.

Preamble to Enclosure Act for Cumberworth and Cumberworth Half

This petition in the preamble to the Act is interesting for the number of parties concerned in the enclosure. The exact boundaries of Cumberworth Half are difficult to determine today.

Whereas there are within the Manor of Cumberworth with Cumberworth Half in the Parishes of Silkstone and Kirkburton in the West Riding of the County of York several Open Commons, Moors and Waste Grounds;

AND WHEREAS Thomas Richard Beaumont of Bretton . . . is Lord of the Manor of Cumberworth . . . and Owner of the Soil of the said Commons, Moors and Waste Grounds . . . and is also Patron of the Rectorial Chapel of Cumberworth aforesaid and Owner of divers tenements . . . within the said Manor;

AND WHEREAS William Railton Clerk is Rector and Curate of the said Chapel of Cumberworth and . . . is entitled to certain Glebe Land and to a certain proportion of the Great . . . Tythes, and also a Proportion of the Small . . . Tythes . . .;

AND WHEREAS the Governors of the Hospital of Gilbert Earl of Shrewsbury in Sheffield . . . are the Impropriators of the remaining Great Tythes . . . arising within the said Manor;

AND WHEREAS . . . Thomas Wickham, Clerk is Vicar of Kirkburton and as such is entitled to a proportionable Part of the Small . . . Tythes arising in the said Manor;

AND WHEREAS the said William Railton, James Milnes, Thomas Hardy, Jonas Kenyon, William Robinson and Caleb Horn, Gentlemen and several other persons being Owners and Proprietors of Lands . . . within the said Manor . . . being entitled to Right of Common on the said Commons . . .;

AND WHEREAS the said Commons, Moors and Wastes Grounds are in their present Situation incapable of Improvement and it would be advantageous to the several persons interested therein, if the same were divided, allotted and enclosed;

Enclosure Act for Cumberworth, 1799, Barnsley Holgate Grammar School Library.

Commissioner's oath

The following is an example of such an oath. Wm Hill was a commissioner appointed to survey the Eccleshall, Sheffield, enclosure which sparked off a riot in 1793.

I William Hill do swear that I will faithfully, impartially and honestly according to the best of my skill and Judgement execute the trusts imposed in me as a Commissioner by virtue of an Act of Parliament for dividing and inclosing the several open Commons Moors and Waste Grounds within the Manor and Township of Eccleshall in the parish of Sheffield in the West Riding of the County of York without favour or affection to any person or persons whomsoever.

So help me God
Wm Hill

Eccleshall Award MD 1750, Sheffield City Library.

Notice of the preliminary meeting to survey the land

The usual practice was to call a preliminary meeting in order to ascertain the boundaries of the manor concerned. The following is the notice of the meeting for Thurgoland.

Thomas Gee of Little Houghton, Darfield parish, Commissioner gives notice that . . . I shall attend at the House of Mr Dagley, the Rose and Crown Inn in Peniston in the said County of York on Monday next the Fourteenth Day of March at Eleven o'Clock in the forenoon for the purpose of enquiry into the Boundary of the Manor of Thurgoland aforesaid by the examination of witnesses upon Oath Court Rolls and Records . . . [to enable] me to set out, determine and fix the boundaries of the Manor of Thurgoland. . . .

Wharncliffe Muniments, Wh.M.61-1, Sheffield City Library.

Evidence taken at an enquiry concerning a road

There was a dispute about the road between Wortley and Thurgoland as to what part of it was involved in the enclosure act. Note that the surveyor of the highways is also a colliery owner.

1814 Penistone 13 June Timothy Ibbotson of Wortley aged 61 had known the road from Riley House to Laycock's cottage ever since he was ten years of age — at that time he drove his Father's team from the Coal Pitts in Hand Lane Bottom — the Road continued in Thurgoland Lordship until he got to Rolling's House and then he went sometimes into the Lordship of Wortley and some times Thurgoland according to the convenience of the Ground until he got to the fifth mere stone and then he invariable kept on the Thurgoland Side to the Sheffield Road . . . Mr Corbett was a Surveyor of the Highways in that Township (Wortley) 30 years since & had the Colliery in Hand Lane. At that time he called upon Ibbotson to do Statute Duty upon that part of the Road from where the Road at present entered Wortley till the fifth mere stone where the old Road entered Thurgoland Manor again & that the Witness did so Statute Duty as an inhabitant of Wortley & that the same part of the Road was commoned by the Inhabitants at large of Wortley and that no Inhabitants of Thurgoland commoned there.

Wharncliffe Muniments, Wh.M.61-3, Sheffield City Library.

Allotment from an enclosure to build a school

The enclosure award for Shelley gives details of an endowment for a school in this Pennine village at that time within the parish of Kirkburton.

Three Acres shall, from and after a School House shall be erected at the Expence of the Inhabitants of the said Township of Shelley . . . and a Schoolmaster appointed for the same, be forever hereafter appropriated for the Benefit of the said Schoolmaster. . . . Three other Acres shall, from and after the said School House be erected and a Schoolmaster appointed to the same be forever thereafter appropriated for the like Benefit of the said Schoolmaster for the time being. . . .

Shelley Enclosure Award, 1799, Barnsley Holgate Grammar School Library.

document 39

Commutation of tithes at enclosure

The enclosure of Cumberworth extinguished the modus in lieu of tithe and replaced this by allotments of land as compensation.

A Modus of Two Pounds and Four Shillings is or are Annually payable to the Trustees of Sheffield Hospital of which Two Shillings is . . . in lieu of Tythe Hay out of a Part . . . of the Hamlet of Cumberworth Half and Two Pounds Two Shillings is Payable Annually to the said Trustees in lieu of Tythe Hay, Corn and all other Grain, Predial and Rectorial Tythe out of the Township of Shepley . . . from the passing of this Act the sum of Two Pounds Two Shillings only be payable in lieu of Tythes to the said Trustees. . . .
That the said Commissioner shall . . . allot and award unto the Rector of Cumberworth, the Vicar of Kirkburton and the Trustees of Sheffield Hospital and their successors (over and above the several Allotments . . . in lieu of their respective Tythes of the old enclosed Grounds . . ., one Tenth part of the Residue of the said Commons, Moors and Waste Grounds and no more in lieu of and as a Recompense and full Satisfaction for all their and each and every of their Great and Small Tythes, Moduses, Compositions and other Ecclesiastical Payments . . . arising and renewing and increasing . . . or be claimed out of any or all of the Lands or Grounds hereby intended to be divided and inclosed.

Cumberworth Enclosure Act, 1799, Barnsley Holgate School.

Claim for an allotment at Birdwell

I James Saville of Birdwell in the Township of Worsborough do hereby claim a Right of Common & an Allotment in respect thereof for one Messuage or Tenement in my own Occupation and situate at Birdwell aforesaid the same being a Freehold Estate of Inheritance.

> Dated this 13 Day of June 1817
> James X [his mark] Saville

The appointed Claim to Common Right is Jane Saville. This is disallowed 3 October 1817 Howard Gee.

Thomas Spooner, Enclosure Commissioners' Minute Book for Worsborough, 1817, p.74, NBC/91, Sheffield City Library.

Thurgoland enclosure – Instructions for fencing

The Commissioners laid down detailed specifications for fencing.

If you wish to enclose with Posts, Rails & Quickwood, the Quickwood to be planted not less than 4 feet from the Line of Stakes & the Ditches to be made 3 feet wide at the top or if with Stone Walls such Walls to be made not less than 5 feet high. . . .

> W. Bingley

Wharncliffe Muniments, Inclosure Papers, Wh.M.61-5.

Enclosure at Barnsley in 1777

In this case full details are given concerning the preservation of the water supplies for the town following enclosure.

And we doe further direct and award that the Wells and Springs called Beverhole Well, the Oak Well, the Crow Well and the Shaw Well the Honey Well and the Watering Place on Poggmoor Common are and shall be left open for the benefit of the inhabitants of the said township of Barnsley. And whereas for the better accomodation of the said Inhabitants we have conducted the Well water from the Warren Well and other springs arising in the allotments marked No. 862 by a covered drain to a stone trough now fixed at the North East Corner of the same Allotment where the Sheffield and Wakefield Turnpike Road crossed the Doncaster and Saltersbrook Turnpike Road, we do order, direct and award that the Constable of Barnsley aforesaid and their successors for the time being or their workmen shall and lawfully may at all seasonable times hereafter when necessary enter into and upon the same allotment and take up, cleanse and repair the said covered drain.

Barnsley Enclosure Award, 1777, NBC/68, Sheffield City Library.

document 43

Problem of cultivating enclosed land at Arlecdon, 1821

A letter from the Vicar of Arlecdon to the Commissioner concerning problems involved in breaking up the ground.

Whelpside
Sept. 20 1821

Sir,
    The Allotment of Common made to the Lord Bishop of Chester from the Inclosures of the Wastes of Arlecdon, Frizington and Whillimoor amounts altogether to 437 Acres. About 100 Acres of which cannot without a very heavy expense be brought under Cultivation, some of the Remainder may be made good Ground, the rest indifferent. The Tithes of Frizington may be reckoned at £75 a Year. I intend taking a Gentleman over the common Lands who has had experience in breaking up such Grounds in order to ascertain as just a Rental of these Grounds as I can obtain from him and will forthwith send you his value of each separate Allotment.

I have no doubt but the Rectorial Value of the Living of Arlecdon will be very considerably increased by the Division of the Commons. You will hear from me on this Head very shortly. . . .

Your most obdt Servt
Jos Fullerton

Arlecdon Parish Papers, DRC/10, Carlisle Record Office.

document 44
Sentence following an enclosure riot, 1815

The following is an example of a vicious sentence passed upon one person who dared to speak up in Court for the rights of the poor after the riot at Lopham in Norfolk.

Wm Mason, Thomas Brook, Edmund Chilley and Ann Rush were indicted for wilfully and maliciously damaging a fence the property of Charles Green, made under the Lopham Inclosure Act which offence is made felony by Act of Parliament. The prisoner Mason addressed the Court as champion of the rights of the poor, whose property he said the Commons were. The Judge stated distinctly to the prisoners and the Jury that the poor had no such right as was asserted by the prisoner Mason. In the present case it appeared that 200 acres of land had been allotted for the use of the poor. Mason was sentenced to twelve and the others to three months imprisonment.

W. E. Tate, private collection.

document 45
Comment on wealthy farmers

Arthur Young had little sympathy for those farmers who acquired wealth by enclosure and used it in trying to ape the aristocracy.

I see sometimes, for instance, a pianoforte in a farmer's parlour which I always wish was burnt; a livery servant is sometimes found and a post chaise to carry their daughters to assemblies. Those ladies are sometimes educated at expensive boarding schools, and the sons often at the University to be made parsons but all these things imply a departure from that line which separates these different orders of beings. Let all those things and the folly, frippery, foppery, expense and anxiety that belongs to them remain among gentlemen.

Arthur Young, *Report on Agriculture*, 1795.

document 46

Enquiry into supplies and prices of food, 1799

Conditions were so bad that the Board of Agriculture sent out these questions in order to discover what conditions were throughout the country.

1 What is supposed to be the stock of wheat and rye in hand relative to the consumption of the remainder of the year, previous to the next crop coming on the market?
2 What are the expectations of next year's supply relative to any deficiency which it is supposed may result from the autumnal rains and the present severe frost?
3 What have been the most successful methods adopted for the relief of the poor?
4 What has been the rise if any, in the pay of agricultural labour, on comparison with preceding periods?
5 Has any article of food, as a substitute for wheaten bread, been successfully used?
6 What is the present price per pound of mutton, beef, pork, butter, cheese, potatoes regard being paid to such joints of meat as come within the consumption of the poor?
7 What is the present and ordinary price of coals?
8 What has been the effect of the frost on turnips, cabbages, and other articles of green winter food for cattle and sheep, also on the young wheat?
9 What is the present price of straw and hay per ton?
10 What is the present or late price of the wool of your country?

Annals Of Agriculture, vol. 24, 1795.

Summary of harvest deficiencies 1794-1800

The following is a shortened version of the reports upon the seasons.

1794   Harvest generally deficient.
1795   A very bad winter. Sharp frosts in late June that thousands of new shorn sheep died of extreme cold. Harvest a fifth or quarter less than normal.
1796   Harvest poor.
1797   The summer was the wettest in living memory and following winter wet with outbreak of sheep rot.
1798   A poor autumn with severe early frosts and heavy falls of snow in the winter followed by rapid thaws which ruined much of turnip crop.
1799   Harvest spoilt by heavy rain and early frosts with an unfavourable autumn.
1800   Early frosts with snow and rapid thaws and a wet spring. Harvest failed.

J. D. Chambers & G. E. Mingay (21); E. L. Jones (38).

Enquiry into food supplies and prices, 1800

The Duke of Portland sent out a circular to all the bishops asking them to get their parochial clergy to obtain the information required as being the quickest and most reliable method of getting details.

1 What has been the produce of the late Crop in your district and county, so far as you have means of information comparative with former crops, or if you can state it, what is the estimated number of Bushels per acre distinguishing Wheat, Barley, Oats and Potatoes and also the comparative Crops of Hay, Beans and Turnips?
2 What is the Price in your Market Town of Wheat, Barley, Oates, Potatoes, Hay, Beans and Mutton and Beef in the month of October 1800 comparatively with the same period in 1799 and 1798?

3 Has the produce of the late Harvest been consumed? Is there any reason to believe that there is any considerable quantity of old wheat in store?
4 Has much foreign wheat or Flour been bought into your part of the country? Is there much use made of Rice Barley or Oats as substitutes?

Home Office Papers, HO.42, 52, P.R.O.

document 49

Bread rationing, 1801

The amount of bread per person was specified and the use of flour and oats severely restricted as the result of a series of bad harvests.

Our Royal Proclamation recommending to all such Persons as have the Means of procuring other Articles of Food the greatest Economy and Frugality in the Use of every Species of Grain: We, having taken the said Address into Consideration and being persuaded that the Prevention of all unnecessary Consumption of Corn will furnish One of the surest and most effectual Means of alleviating the present Pressure and of providing for the necessary Demands of the Year have . . . thought fit . . . to issue this Our Royal Proclamation, most earnestly exhorting and charging all those of Our Loving Subjects who have the Means of procuring other Articles of Food than Corn . . . to practise the greatest Economy and Frugality in the Use of every Species of Grain: And We do for this Purpose more particularly exhort and charge all Masters of Families to reduce the Consumption of Bread in their respective families by at least One Third of the quantity consumed in ordinary Times, and in no Case to suffer the same to exceed One Quartern Loaf for each Person in each Week; to abstain from the Use of Flour in Pastry, and moreover to restrict the Use thereof in all other Articles than Bread; And do also in like Manner, exhort and charge all Persons who keep Horses, especially Horses for Pleasure, so far as their respective Circumstances will admit, carefully to restrict the Consumption of Oats or other Grain for the

Subsistence of the same. And We do hereby further charge
and command every Minister in his respective Parish Church
or Chapel within the Kingdom of Great Britain to read.
... Our said Proclamation on the Lord's Day for Two successive Weeks after receiving the said Proclamation.

Copy in the Archdeaconry of Richmond Archives, ARR/10/2, Lancashire Record Office.

document 50

Farm stock prices in south Yorkshire, 1825

*These prices for fat stock should be compared with those quoted by
Cobbett in his* Rural Rides *i, 13-14. These samples are taken from the
Elmhirst accounts at Worsborough and the Dyson accounts at Winter-
hill Farm.*

Nov. 2   1825   Received at Wakefield £26; 8; 0d for 12
                  ewes at £2; 4; 0d each.
                  Received at Wakefield £10; 16; 0d for 4
                  wethers at £2; 14; 0d each.

Elmhirst Accounts, EM.1187, Sheffield City Library.

Feb. 10	1825	24 stone of bacon at 9d per lb	£10; 16; 0d
Feb. 19		Received for Fowls	7; 0d
Feb. 20		Received for Butter	4; 0; 3d
March 8		Shoulder of Pig	2; 2; 0d
May 26		Mr Hudson for Veal Calf	2; 3; 0d

Winterhill Farm Accounts, MD, Yorkshire Archaeological Society.

document 51

Farm rents

*Part of a letter from Richard Fenton, steward to the Marquis of
Rockingham, concerning the letting of farms, 18th February 1769.*

The farm of £62 a year at Ackworth Moor Top on which James Wood dyed not much above a year ago I let at Candlemas 1768 to his Nephew, George Goodyear at £90 a year clear. . . . Yesterday I attended Mr Townend's sale at Houghton but nobody bid a farthing at the Hoyland Estate.

Wombwell Woodhead was sold to one Thirlwall a tenant of Lord Strafford at £3,300 and it is now let at £80 a year and racked to the utmost. . . .

Rockingham Papers, R.187-27, Sheffield City Library.

document 52

Labour costs and produce from Woodnook Farm, 1771

This extract from the accounts of the Marquis of Rockingham for this farm on his estate should be compared with the capital investment at Blackmoor Farm on the Wharncliffe estates in 1790.

John Willcock for hedging 36 acres at 1s. 2d	£2; 2; 0d
John Willcock for Cuting of hedgwood and Sheaping stacks	4s 0d
Thomas Swift filling and Spred 314 lood of Manure at 1d	£1; 6; 2d
John Grimshaw threshing 24 load of beans at 6d load	12s 0d
John Grimshaw threshing 16 quarters Oats at 9d quarter	12s 0d
John Grimshaw threshing 13 load Whet at 10d load	10s 10d

PRODUCE

Butter	2,229 lbs at 7d lb.	£65; 0; 3d
Cream to the House	1,619 pints at 3d pint	20; 4 9d
New Milk	1,081 quarts at 1d	4; 10; 1d
Pork	142 stones at 4/8d	33 5 8d
22 calves killed	86 stones 6 lbs at 33d	15 2 6d
6 calves brought up		18 0 0d
		£156 3 3d

Rockingham Papers, R.187, 39b, 36, Sheffield City Library.

Blackmoor Farm valuation, 1790

*The following extract from the accounts shows the amount of capital
investment in seeds and fertiliser for one year.*

A.R.P.		£	s	d	£	s	d
3 0 0	Intack in Grass 9 quart Hay seeds Sown	3	7	6			
	36 Loads of Manure half tillage	5	8	0			
					8	15	6
7 0 0	Lindley Field 21 quart of Hayseeds Sown	7	17	6			
	In do 12 Loads of Manure half Till	1	16	0			
					9	13	6
8 0 0	Ing 2 Acres Fallow 4 times Dressed	2	16	0			
	3 Loads of Manure		18	0			
	Rent and Taxes	1	0	0			
	2 Acres Fallowed in 1788 half Dress'd 5 times	1	15	0			
	28 Loads of Manure half tillage	4	4	0			
	32 lbs White Clover Seed Sown	1	1	4			
	Low part of Close sown with Hay Seeds	1	10	0			
	10 Loads of Manure half Tillage	1	10	0			
					14	14	4
7 0 0	Ruff Holm 3 Acres Pair'd not Burn'd	1	10	0			
	2 Acres Pair'd, Burn'd and Plowed part Turnip	2	2	0			
	2 Acres Pair'd, Burn'd and Crop of Oats	1	0	0			
					4	10	0
4 0 0	Dam Holm 12 qrs Hayseeds Sown				4	10	0
	Round about Holm 14 Loads Manure				2	2	0
	Barley Holm 30 Metts Bone Dust	1	10	0			
	In do 30 Loads Manure half Tillage	4	10	0			
					6	0	0
7 0 0	Low Oxspring Field 28 quarters Hay seeds	10	10	0			
	In do 2 Chaldrons of Lime half Tillage	1	2	0			
					11	12	0
1 1 0	Laith Croft 14 Loads Manure half Tillage	2	2	0			
	Plow'd ready to sow with Corn at present		7	6			
					2	9	6

Blackmoor Farm Accounts, Wh.M.30.25, Wharncliffe Muniments,
Sheffield City Library.

Details of tithes from Richmond Terriers

Each year the churchwardens were required to report to the Bishop upon the state and extent of the church property, the source of the benefice stipend and other details. These documents are known as Terriers.

BARMINGHAM 1701 To this Rectorie belongs 22 pasture gates and a half in the low close and 2 gates in the Stale Knowl. The Lordship of Scargill pretends a modus of 40s yearly in lieu of tithe Corne & hay but this is in dispute. The hamlet of Hope pays a composition of eight shillings and two pence for tithe hay & Corne. All other Tythe are paid in kind throughout the whole parish.

BEDALE 1760 All manner of Great Tithes are paid in kind or compounded for throughout the Parish except the Composition for Cowling due at Easter £2; 5s; 0d.

KIRKLINTON 1778 No Common Fields or Right of Common belonging to Kirklington, the Tythes are Paid in kind as it is a Rectory. Only a modus of One Pound Ten Shillings from Upsland Estate.

Richmond Glebe Terriers, RD/G, Nos 4, 7, 35, Leeds City Library.

Enclosure of Barugh Common, 1714

This enclosure took place under the Act of 12 Anne which allowed the churchwardens of any Yorkshire parish to enclose up to sixty acres of common to augment the benefice stipend where it was less than £40 per annum.

Whereas the parish of Darton is of great extent and very populous and hath large wastes of Commons within it. Part of which doth lay in the Manor of Barugh and the Vicar of Darton hath not yet a settled provision of Forty Pounds per annum for his maintainence. Now in pursuance of an Act of Parliament made in the twelfth year of our Late Sovereign Lady Queen Anne. . . . We who have hereunto subscribed our names being the Lords of the Manor and three parts of four of all the freeholders and others who have rights of Common . . . within the Manor of Barugh do here give consent to David Ellison, John Travis, Antony Webster and John Leech to enclose so much of the said Commons as shall not exceed 60 acres . . . and we do agree that the said David Ellison, John Travis, Antony Webster and John Leech be seized of the Wastes or Commons when enclosed . . . (*signed*) John Wentworth, George Marsh, Joseph Spencer, George Beaumont, Nathan Burley, Abraham Barber.

Bretton Hall Muniments, DD70/46, Yorkshire Archaeological Society.

document 56

Popular reaction to enclosure of commons, 1753

This is an extract of two verses from a scurrilous poem concerning the enclosure of Charnwood Forest.

So God bless King George and defend us from Evil
And send all Encroachers on Commons to th' Devil
Let him flea ye sharp Squire who worrys ye poor
Like ours to wax wealthy and live with an whore
Derry down etc.

If Gentlemen venture Damnation at last
Let all honest Fellows be modest and chaste
Let us kiss our Wives, And let Wenches alone
And every man be content with his own
Derry down etc.

W. E. Tate, private collection.

document 57

Hygiene in the sale of milk, 1779

This extract shows that hygiene played little part in the sale of food-stuffs in towns. It is therefore no surprise that the incidence of disease was very high.

Milk is carried through the streets in open pails, exposed to the foul rinsings discharged from doors, windows, spittle, snot and tobacco quids, from foot passengers; overflowings from muck carts, spatterings from coach wheels, dirt and trash chucked into it by roguish boys for the joke's sake, the spewing of infants, who have slobbered into the tin measure which is thrown back in that condition amongst the milk for the benefit of the next customer; and finally the vermin that drops from the rags of the nasty drab that vends this precious mixture under the respectable denomination of milk maid.

Tobias Smollett, *The Expedition of Humphrey Clinker*, 1770.

document 58
Report on the state of Yorkshire crop returns 1801

These replies for the parishes of Fishlake, Bradfield and Penistone are in reply to the questionnaire sent out by the Duke of Portland, to the bishops in 1800, asking that the parish clergy be requested to report on agriculture in their parishes [doc. 48].

FISHLAKE; The Land of this Parish, a small part excepted, is a strong clay, on a wet bottom, and is found very un-productive in a wet season — it proved particularly so in the years 1799 and 1800.

BRADFIELD; I ... called public meetings of the Inhabitants and spent several days in going amongst the Farmers; still for want of information I have been obliged to compute the quantity of Acres of Corn, in several farms, from the quantity of Acres in the whole far.

PENISTONE is very extensive yet there is little Land upon the Plow in Comparison, the Farms in general being graz-ing and Stock Farms.

D. G. Hey (68).

document 59
Report by William Martin to Earl Fitzwilliam, 1770

Estate stewards maintained a regular correspondence with the owners keeping them informed of progress of the crops, the expected yield and other estate matters.

My Lord,

Inclosed your lordship will receive the best account I can at present produce of the farm at Woodnook, for the year 1769. You will please to observe that we have computed the quantity of Wheat and beans remaining unthreshed; as soon as they are brought in, I will produce you an exact Account and then this I now send may be destroyed. Under another cover I send by the same post an Account of the Street Farm, which is also imperfect by reason of part of the crop being unthresh'd, however this will soon be finished.... There does not seem to be any great prospect at present in either of these farms, especially the Beans. There will not be half a crop in any part of them. Welland will have some tolerable Wheat and Barley. The Lucerne at Street is good for nothing. The weather here is wet and very cold we have not yet scarce had the appearance of summer. The crop of Hay will not be great. Your Lordship's directions about the House building shall be immediately put into execution.

Inclosed you have an account of the produce of the arable land, lately taken into hand at Tankersley. The Wheat growing there on the land dress'd with salt promises greatly but that at Street, it seems to have had no effect upon. We have had another great amount of rain for several days so that the roots are exceeding bad and the ground very wet. They are just now going to bury John Mann.

> I am my Lord
> Your Worships most obedient servant
> William Martin

Wentworth Woodhouse Estate Correspondence, R.187-1,7.

document 60

Extracts from the Earl of Harewood's estate correspondence

The letter books of the Harewood estates contain a great deal of information since the steward preserved copies of his own letters to the Earl and those he received from the Earl.

14 May 1765; . . . I do not think you made a bad market for your seed, your white & red Clover seed is a penny p. pound under our Markets. Endeed the hay seed are 2d p. Quarter higher but if they prove good growers they may be better by twice that Sum for nothing can be worse than bad hay seeds. 12 July 1765; . . . We have had a strong Drought in this part which still continues & not more than one third of a Crop of Hay, the Pasture Grounds suffer equally; We have got about half of our Hay very dry & are mowing the remainder. Wheat in general looks well but Oats, Beans, Barley will be very short. . . .

Stewards' Letter Book to Earl of Harewood, Harewood Estate Papers, Leeds City Archives.

document 61

Letter to the Marquis of Rockingham about Knottingley lime

The following is an extract from a long letter to the Marquis of Rockingham by John Paine of Newhall of the results of experiments in liming soil.

It falls greater quantity than our Country lime and seems of a milder quality. 15 dozens were laid upon Eight Acres of Light Land fallow for Turnips with forty loads of year old Dung, but to make the experiment more compleate one acre was manured wholly with it (three dozens) the Turnips are of the first rank in the Neighbourhood, those limed and dung'd no better than those lim'd only. Am informed its effects are not so visible in the first as in the six succeeding years.

Having heard it was not so beneficial upon clay as the lime of this country, I resolved to make the experiment and have spread three dozens of the one & two of the other upon two contiguous Acres (it is known a greater quantity than two of ours would be prejudicial). It was Laid on previous to its second plowing and worked in with the successive ones, the remainder of the Close was dung'd with the usual quantity. As it was only sown at Michaelmas can yet say nothing of the effect.

One peculiarity very visible in this Lime is the Turnips in the place where the heap stood before spread are the most flourishing — had it been our Lime it wou'd have produced nothing but Weeds.

I am
J. Payne jnr.

Rockingham Correspondence, R.187-41, Sheffield City Library.

document 62

Winterhill Farm Accounts, 1819

These extracts are selected to show the part played by country farmers in industrial regions in the development of industry.

January 15	1 Load of Coal Hoyland		1s	9d
Feb. 8	1 Load of Coal Holling House		2s	8d
Feb. 11	1 Half day leading Stones for roads		1s	6d
March 13	1 Load of coals Holling House		2s	8d
May 23	1 Load of wood Bultcliff to Hoyland		12s	0d
July 27	3 Horses 1 Day leading furnace hearth		6s	0d
August 16	1 Load of Lime from Emsel (Elmsall)	£1	1s	0d

Winterhill Farm Accounts, High Hoyland, MD, Yorkshire Archaeological Society.

Bibliography

PRIMARY SOURCES

Yorkshire Deeds, vols. i-x, Yorkshire Archaeological Society Records Series.

Duchy of Lancaster, Miscellaneous Books, P.R.O.

Home Office Papers, P.R.O.

Bretton Hall Archives, Yorkshire Archaeological Society.

Earl of Mexborough's Archives, Leeds City Library.

Vavasour Archives, Leeds City Library.

Tithe Cause Papers, Borthwick Institute, York.

Wentworth Woodhouse Estate Papers, Sheffield City Library.

Marquis of Rockingham's Papers, Sheffield City Library.

Scarisbrick Hall Archives, Lancashire Record Office.

Egton Enclosure Award, Lancashire Record Office.

Archdeacon of York's Court Book, 1723, Borthwick Institute, York.

Elmhirst Farm Accounts, Sheffield City Library.

Earl of Wharncliffe's Archives, Sheffield City Library.

Spencer Stanhope Archives, Sheffield City Library.

Cumberworth and Shelley Enclosure Act, Barnsley Holgate Grammar School.

Newman and Bond, Sheffield City Library.

Barnsley Enclosure Award, Sheffield City Library.

Parish Papers for Copeland Deanery, Carlisle Record Office.

Archdeaconry of Richmond Archives, Western Deaneries, Lancashire Record Office.

Archdeaconry of Richmond Archives, Eastern Deaneries, Leeds City Library.

Winterhill Farm Accounts, John Addy, private collection; micro-film in Reading University.

Earl of Harewood's Estate Papers, Leeds City Library.

Parish Papers, Chester Diocese, Cheshire Record Office.

PRIMARY PRINTED SOURCES

1 Cobbett, W., *Rural Rides, 1830*, Dent, Everyman.

2 Defoe, D., *A Tour through England and Wales, 1727*, Dent, Everyman.

3 Fitzherbert, J., *Boke of Husbandrye, 1523.*
4 Marshall, W., *Reports to the Board of Agriculture, 1788-1808*, reprinted, David & Charles, 1969.
5 Marshall, W., ed. *The Review and Abstract of the County Reports to the Board of Agriculture, 1808-18*, David & Charles, 1968, 5 vols.
6 Marshall, W., *Rural Economy of the Midland Counties, 1790*, David & Charles, 1969.
7 Marshall, W., *Rural Economy of Norfolk*, David & Charles.
8 Marshall, W., *Rural Economy of Yorkshire*, David & Charles.
9 Rogers, W. S., 'Parliamentary Enclosure in the West Riding, 1729-1850' (thesis), Brotherton Library, Leeds University.
10 'Shelley Enclosure Act with Cumberworth Half, 1799'.
11 Tusser, T., *Hundreth Good Points of Husbandry, 1557.*
12 Young, A., *An Inquiry into the Propriety of Applying Wastes . . . to the Support of the Poor.*
13 Young, A., *A Six Months Tour through the North of England, 1778*, Dent, Everyman.
14 Young, A., *A Tour through the North of England*, vol. i.
15 Young, A., *Travels in England and France, 1778*, Doubleday, 1969.

GENERAL WORKS

16 Addy, J., *A Coal and Iron Community in the Industrial Revolution,* Longman, 1970.
17 Ashton, T. S., *Economic Fluctuations in England, 1700-1800*, Oxford University Press, 1959.
18 Barnes, D. G., *History of the English Corn Laws*, Routledge, 1930.
19 Beresford, M., *The Lost Villages of England*, Lutterworth Press, 1954.
20 Best, G. F. A., *Temporal Pillars*, Cambridge University Press, 1964.
21 Chambers, J. D. and Mingay, G. E., *The Agricultural Revolution, 1750-1850*, Batsford, 1966.
22 Clayton, A. K., *A Study of Hoyland* (typescript), Sheffield City Library.
23 Clew, K., *Kennet & Avon Canal*, David & Charles, 1968.
24 Eden, F. M., *State of the Poor*, ed. L. Rogers, 1929.
25 Ernle, Lord, *English Farming, Past and Present*, eds. G. E. Fussell and O. R. MacGregor, Heinemann, 1961.
26 Fussell, G. E., *The Farmer's Tools*, Pergamon Press, 1952.
27 Gonner, E. K., *Common Land and Inclosure*, rev. edn., Cass, 1966.
28 Habbakuk, H. J., *The English Land Market in the Eighteenth Century*, Cambridge University Press, 1940.

29 Habbakuk, H. J., *English Landownership, 1680-1740*, Cambridge University Press, 1940.

30 Hammond, J. L. and Hammond, B., *Two Towns Enclosure*, new edn., Longman, 1966.

31 Hammond, J. L. and Hammond, B., *The Village Labourer*, new edn., Longman, 1966.

32 Harris, A., *Open Fields of East Yorkshire*, East Yorks Local History Society, 1956.

33 Hill, C., *Economic Problems of the Church*, Oxford University Press, 1956.

34 Hoskins, W. G., *Leicestershire Farmers in the Seventeenth Century*, Liverpool University Press, 1948.

35 Hoskins, W. G., *Leicestershire Farmers in the Eighteenth Century*, Liverpool University Press, 1950.

36 Hoskins, W. G., *The Making of the English Landscape*, Hodder & Stoughton, 1955.

37 Hoskins, W. G., *The Midland Peasant*, Macmillan, 1965.

38 Jones, E. L., *Seasons and Prices*, Allen & Unwin, 1964.

39 Leonard, E. M., *Enclosure of Common Fields in the Seventeenth Century*, Cass, 1965.

40 McKisak, M., *The Fourteenth Century* (Oxford History of England), Oxford University Press, 1959.

41 Mingay, G. E., *English Landed Society in the Eighteenth Century*, Routledge, 1963.

42 Mingay, G. E., *Size of Farms on the Eighteenth Century*, Routledge, 1963.

43 Outhwaite, R. B., *Inflation in Tudor and Stuart England*, Macmillan, 1969.

44 Parker, C. R., *Enclosures in the Eighteenth Century*, Historical Association, Aids to Teachers Series.

45 Pawson, H. C., *Robert Bakewell*, Murray, 1957.

46 Platt, C. P. S., *Monastic Grange in Medieval England*, Macmillan, 1969.

47 Ramsey, Peter, *Tudor Economic Problems*, Gollancz, 1963.

48 Rolt, L. T. C., *Navigable Waterways*, Longman, 1969.

49 Tate, W. E., *English Village Community and the Enclosure Movement*, Gollancz, 1967.

50 Tawney, R. H., *Religion and the Rise of Capitalism*, Murray, 1926; Penguin, 1969.

51 Thirsk, J., *English Peasant Farming*, Routledge, 1957.

52 Thirsk, J., *Victoria County History; Leicestershire*, vol. i.

53 Thirsk, J., *Tudor Enclosures*, Historical Association, 1959.

54 Thompson, E. P., *Making of the English Working Class*, Gollancz, 1963; Penguin, 1968.

55 Thompson, F. M. L., *English Landed Society in the Nineteenth Century*, Routledge, 1963.

56 Trow-Smith, R., *English Husbandry*, Routledge, 1959.

57 Slicker von Bath, B. H., *Agrarian History of Western Europe, 1500-1850*, Edward Arnold, 1963, Part iii, Section D.

58 Ward, W. R., *Land Tax Assessments in the Eighteenth Century*, Oxford University Press, 1963.

59 Wilkinson, O., *Agricultural Revolution in the East Riding of Yorkshire*, East Yorks Local History Society, 1956.

60 Willan, T. S., *River Navigation in England, 1600-1700*, Oxford University Press; 2nd edn, Cass, 1964.

61 Beresford, M., 'Lost villages of Yorkshire', *Yorkshire Archaeological Society Journal*, vol. 38, pp. 284-86.

62 Bishop, T., 'Assarting', *Essays in Economic History*, vol. ii.

63 Carr, P., 'Open Field Farming', *Derbyshire Archaeological Journal*, vols. 81, 82, 83.

64 Dodd, J. Phillip, 'South Lancashire in transition', *Historical Society of Lancashire and Cheshire*, vol. 117.

65 Fussell, G. E. and Compton, M., 'Agricultural adjustments after the Napoleonic Wars', *Economic History Review*, 1959.

66 Fussell, G. E., 'Low Countries influence on English farming', *English Historical Review*, vol. 74.

67 Fussell, G. E., 'Size of English cattle in the eighteenth century', *Agricultural History*, vol. 3.

68 Hey, D. G., 'The 1801 crop returns for South Yorkshire', *Yorkshire Archaeological Society Journal*, vol. 42, 1971.

69 Holderness, B. A., 'Personal mobility in some rural parishes of Yorkshire, 1772-1822', *Yorkshire Archaeological Society Journal*, vol. 42, 1971.

70 Hoskins, W. G., 'Leicestershire farmers in the sixteenth century', *Agricultural History*, vol. 25.

71 John, A. H., 'Agricultural productivity and economic growth in England', *Journal of Economic History*, vol. 25.

72 John, A. H., 'Course of Agricultural Change, 1660-1760', *Studies in the Industrial Revolution*, 1960.

73 Jones, E. L., 'Agricultural labour market in England, 1793-1872', *Economic History Review*, 2nd series, vol. 17.

74 Jones, E. L., 'Eighteenth century changes in Hampshire Chalkland farming', *Agricultural History Review*, no. 8.

75 Parker, R. A. C., 'Coke of Norfolk and the Agrarian Revolution',

Economic History Review, 2nd series, vol. 3.
76 Rogers, W. S., 'West Riding Commissioners of Enclosures', *Yorkshire Archaeological Society Journal*, vol. 59.
77 Tate, W. E., 'Opposition to parliamentary enclosure in the eighteenth century', *Agricultural History*, vol. 19.
78 Tate, W. E., 'The cost of parliamentary enclosures in England', *Economic History Review*, 2nd series, vol. 5.

GENERAL READING
79 Ashton, T. S., *Economic History of England: Eighteenth Century*, Methuen, 1955.
80 Gaird, Sir J., *English Agriculture, 1850-1851*, 1966.
81 Habbakuk, H. J., *La Disparition du paysan Anglais*, Annales XX, 1965.
82 Haldane, A. R. B., *The Drove Roads of Scotland*, Edinburgh University Press, 1968.
83 Hilton, R. H., *Decline of Serfdom in the Late Middle Ages* (Studies in Economic History), Macmillan, 1969.
84 Hobsbawm, R. and Rudé, G., *Captain Swing*, Laurence & Wishart, 1969.
85 Jones, E. L., *Development of English Agriculture, 1815-1873* (Studies in Economic History), Macmillan, 1968.
86 Mingay, G. E., *Enclosure and the Small Farmer in the Industrial Revolution* (Studies in Economic History), Macmillan, 1968.
87 Minchinton, W. E., 'Agricultural returns and the Government during the Napoleonic Wars' (*Agricultural History Review* i, 1953).
88 Wilson, Charles, *England's Apprenticeship, 1600-1763*, Longman, 1965.

Index

Index